This Book Belongs to

....

Recent studies have shown that adult & kids coloring books can help reduce **stress and promote mindfulness.**

This Coloring Book is a Great Way to Unwind and De-Stress!

We hope you enjoy this product as much as we enjoyed making it!

Happy coloring!

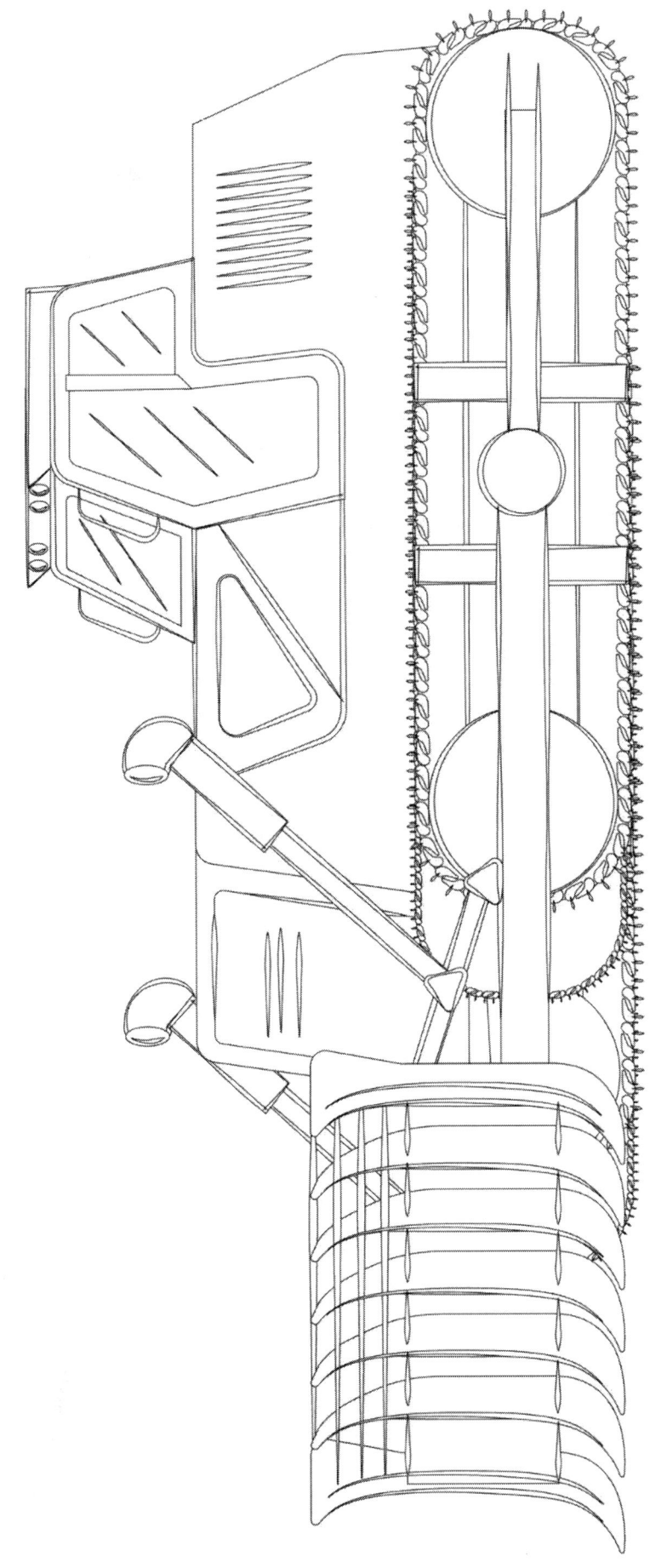

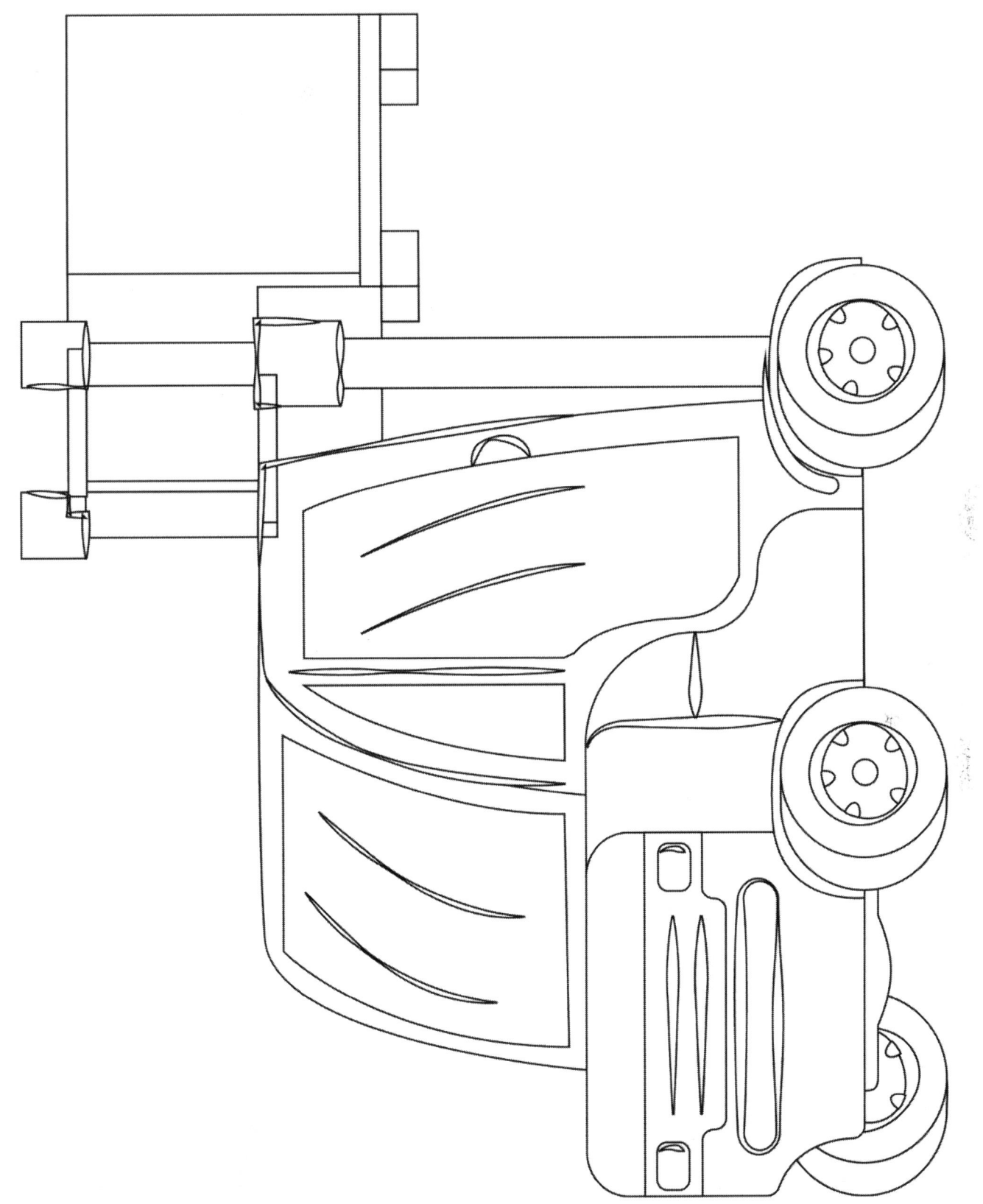

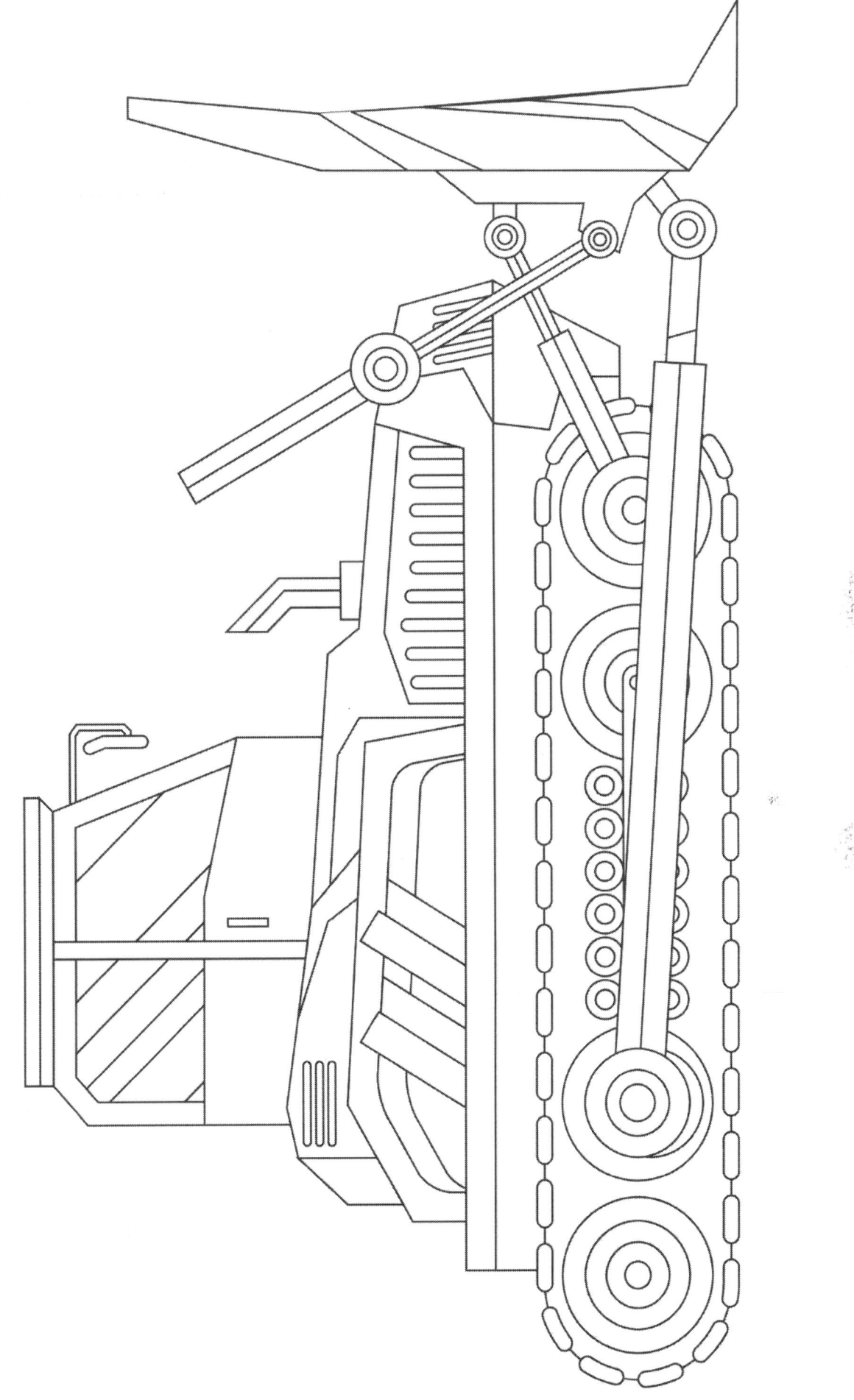

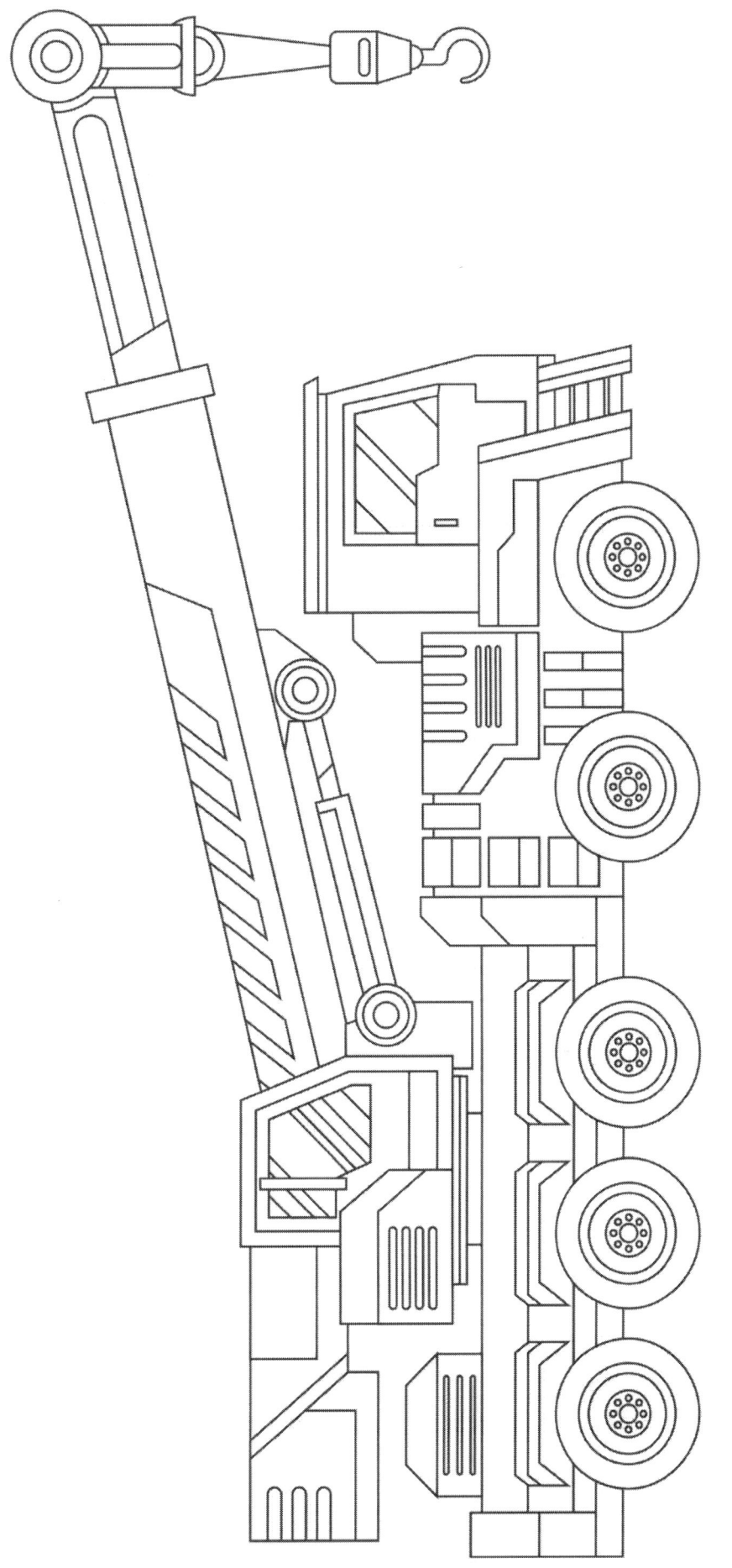

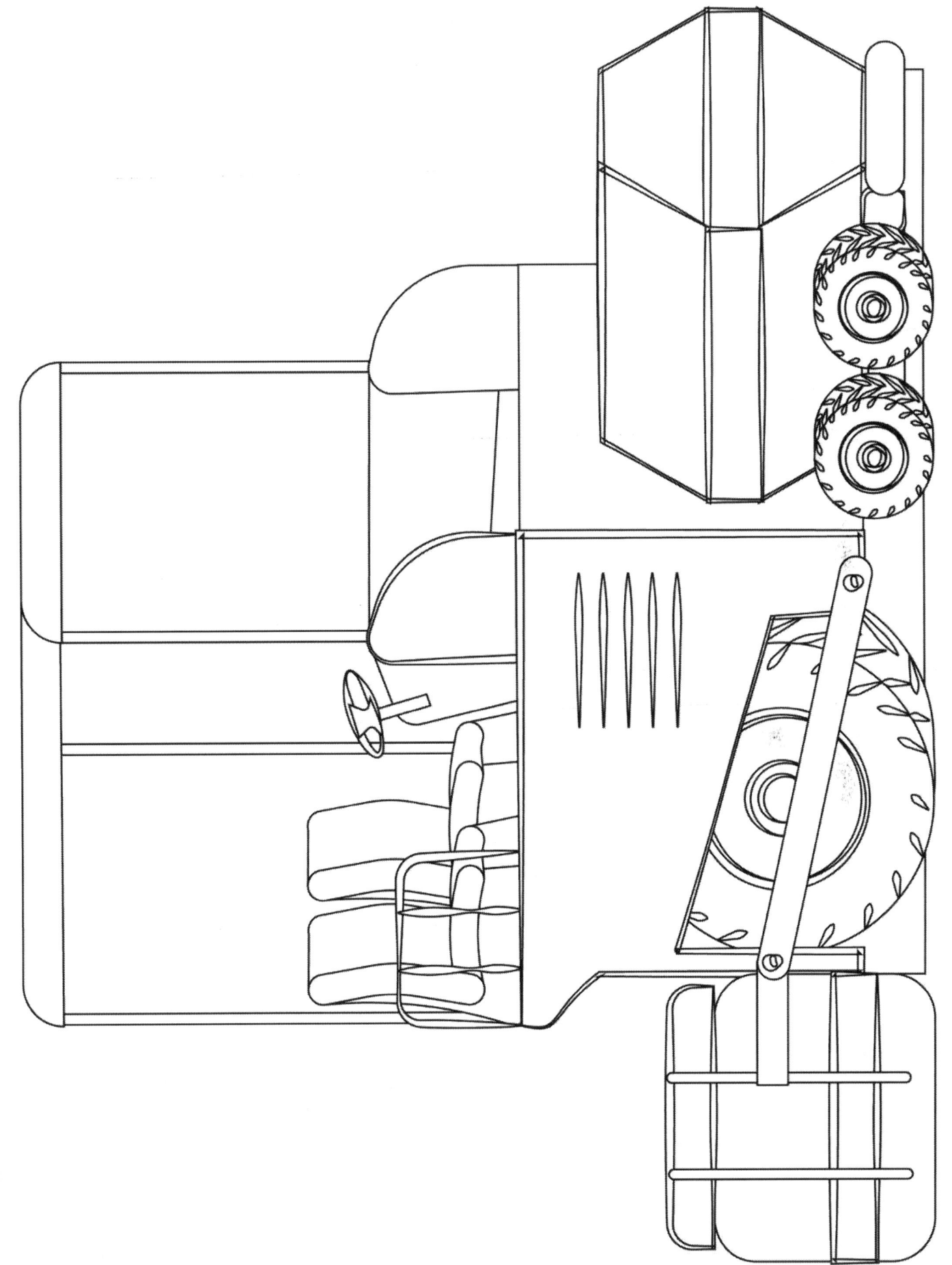

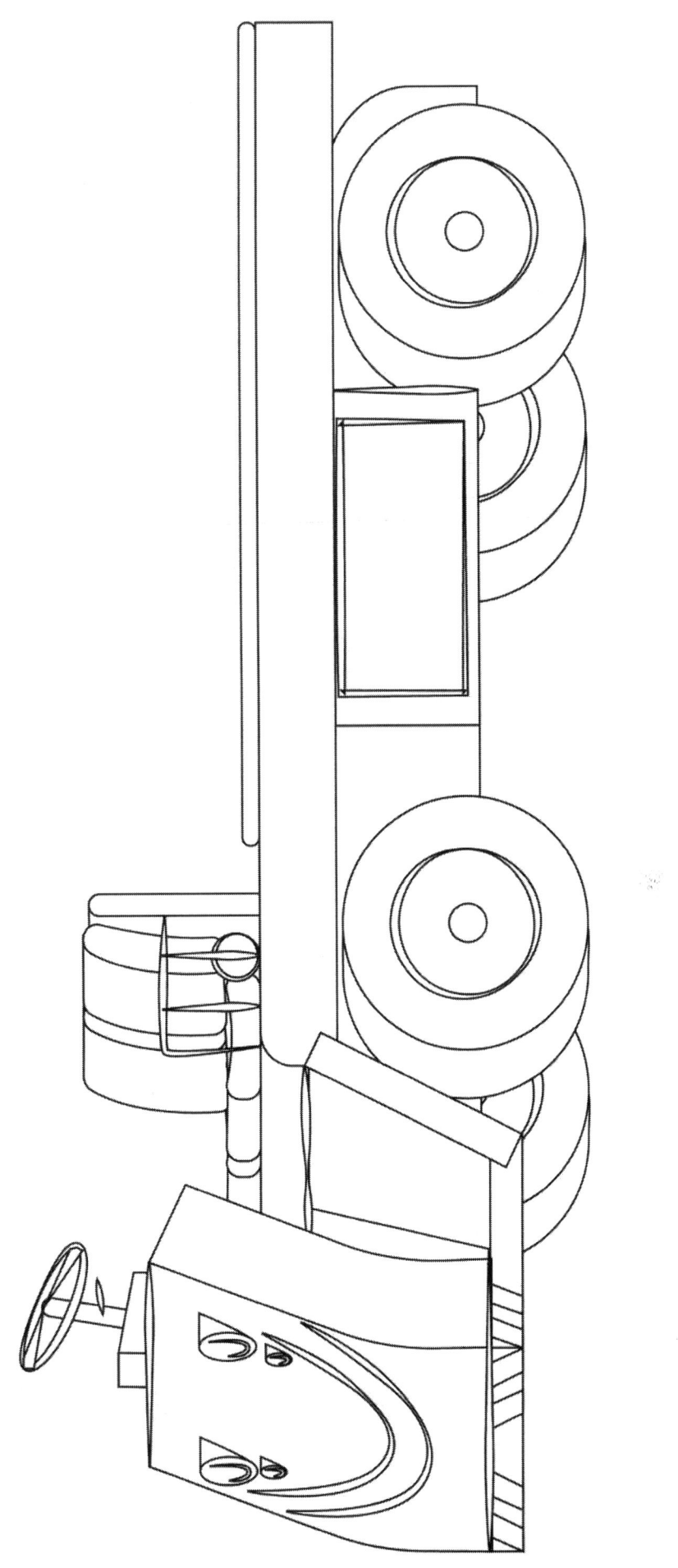

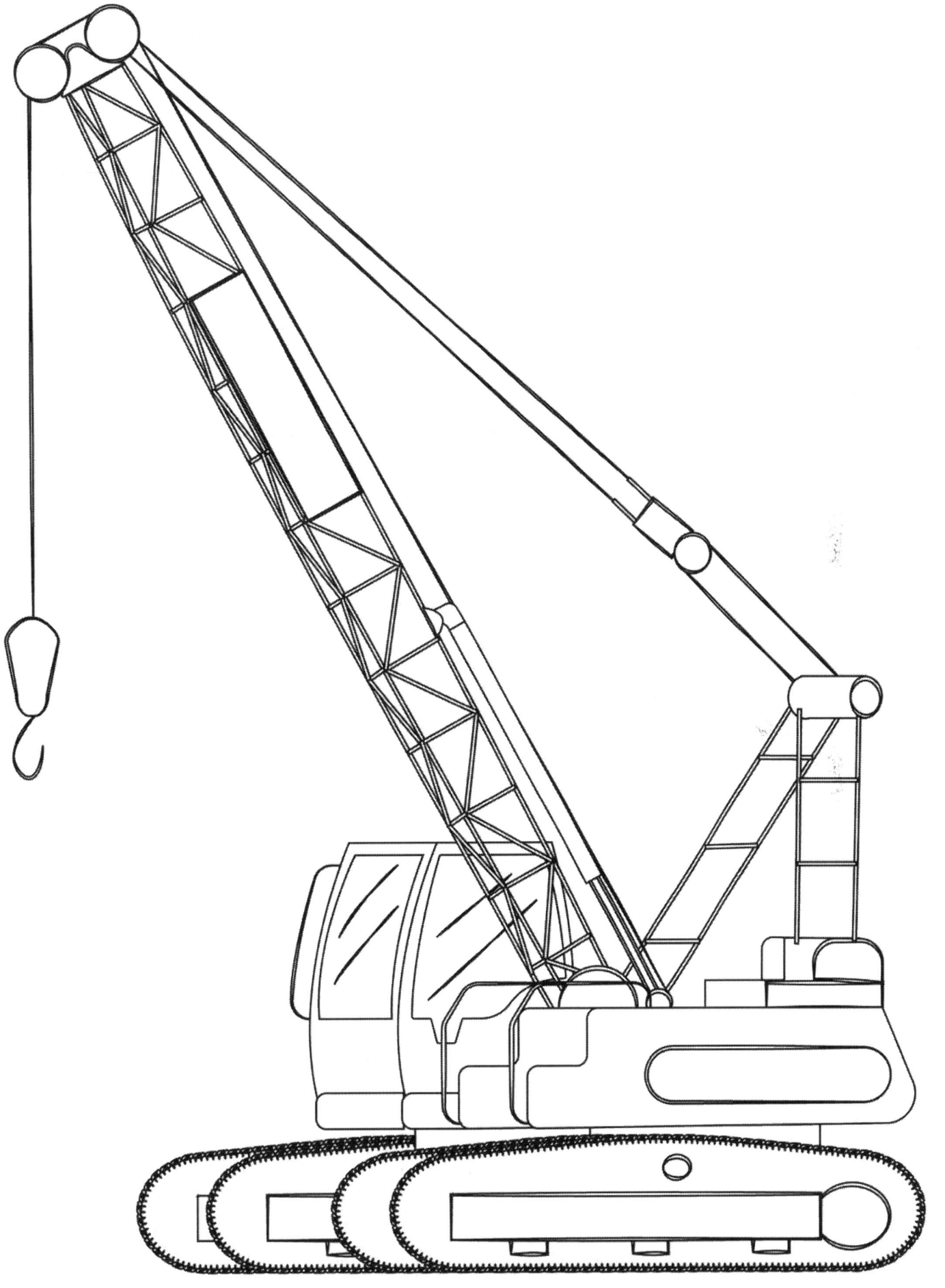

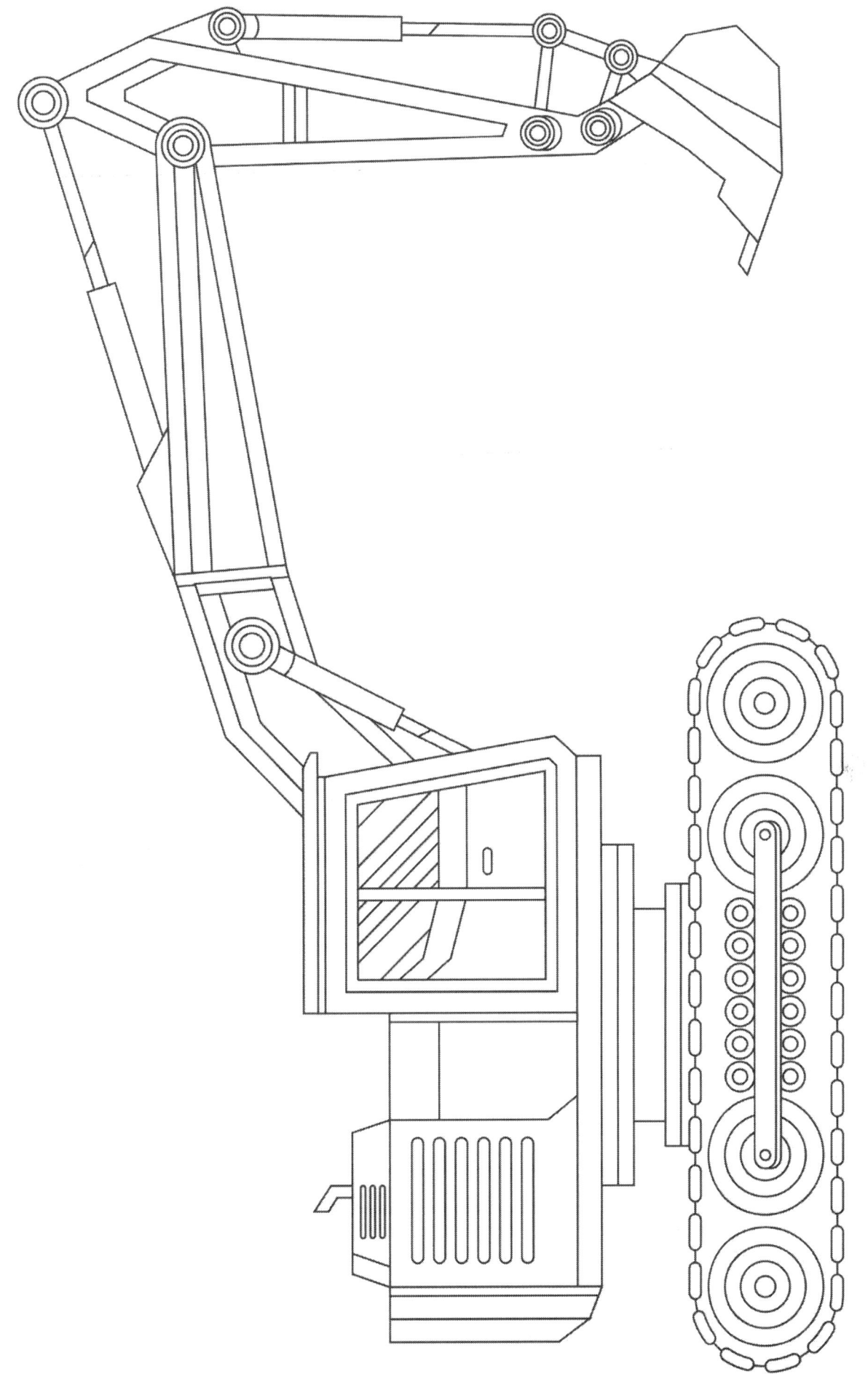

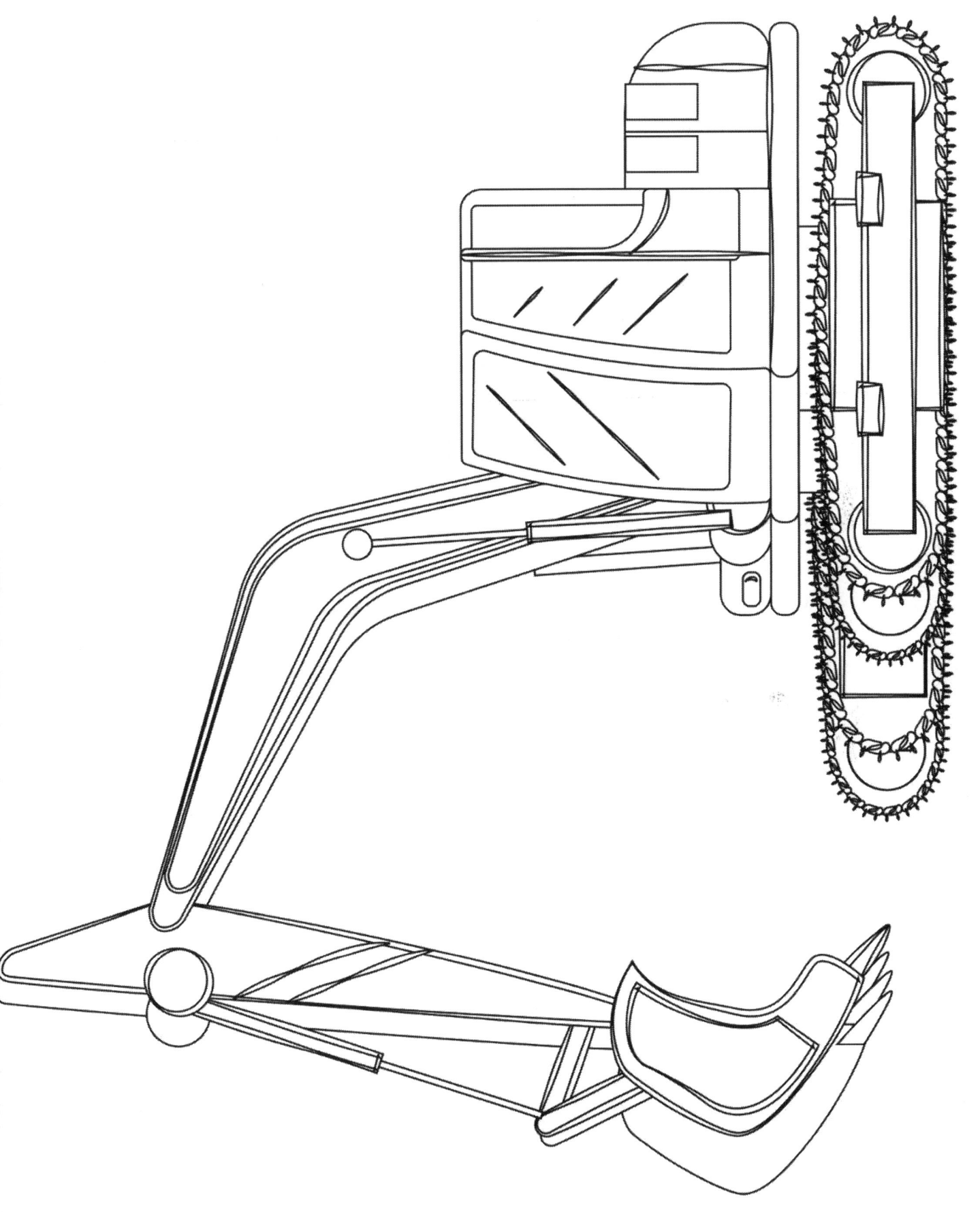

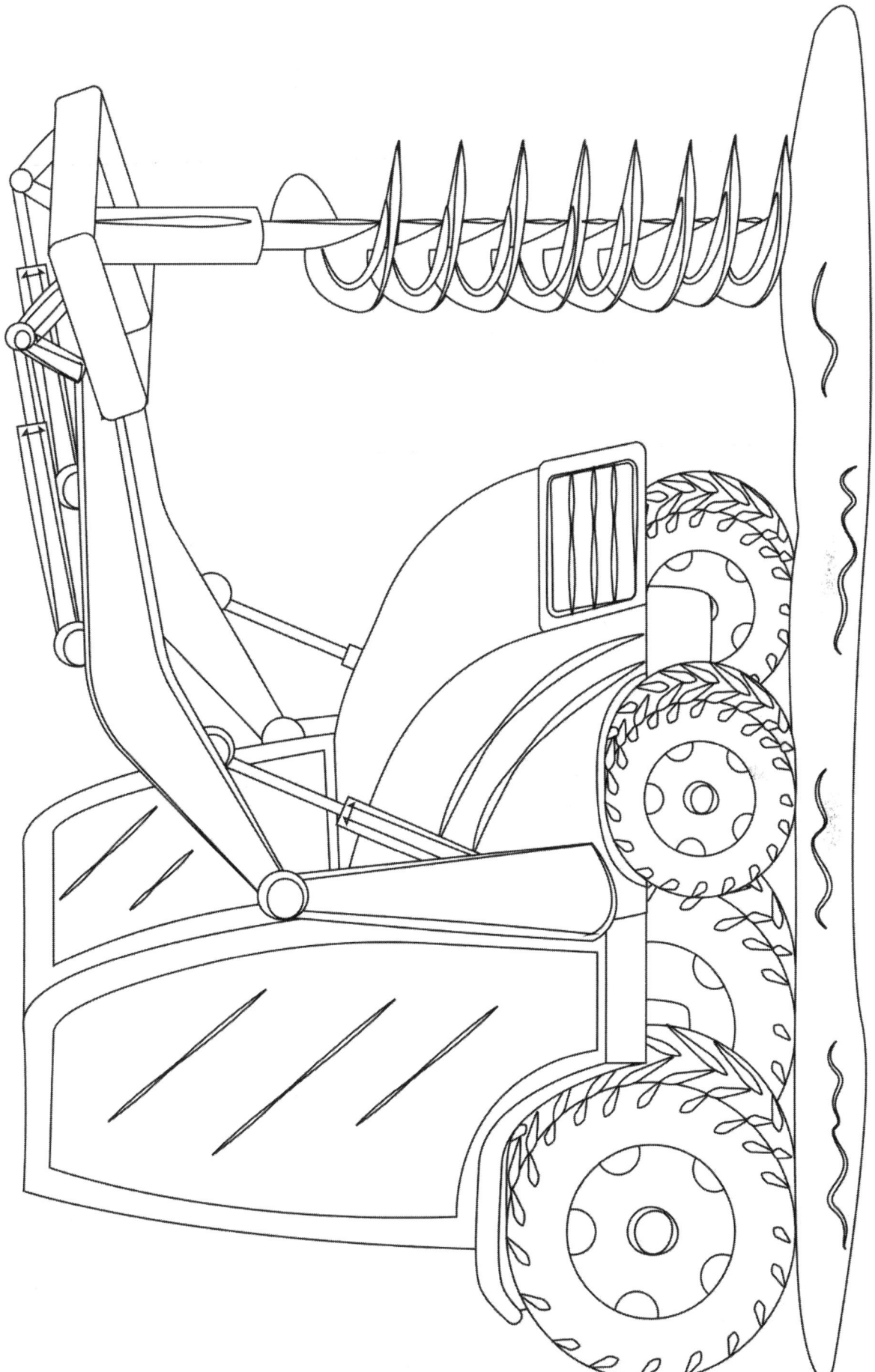

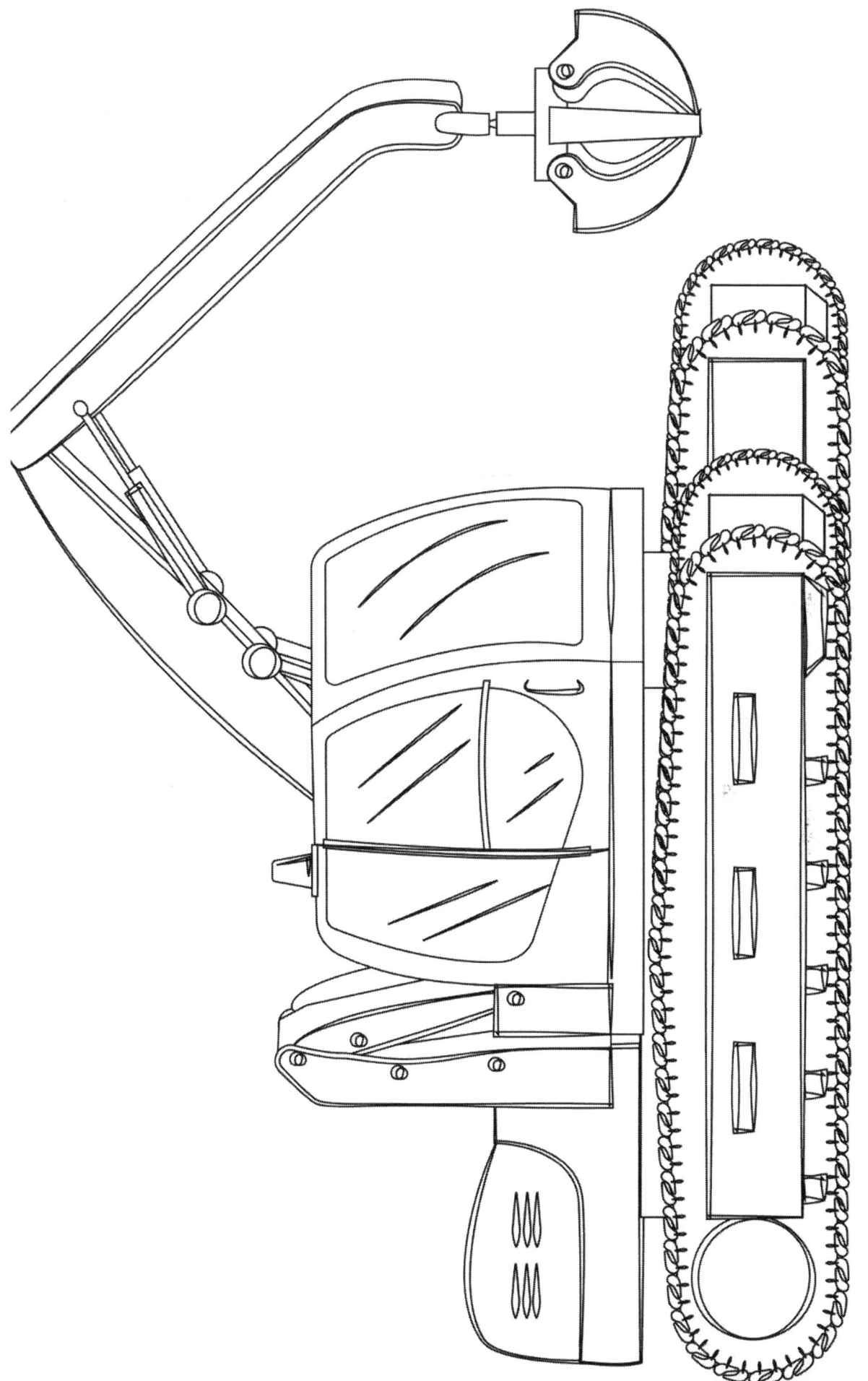

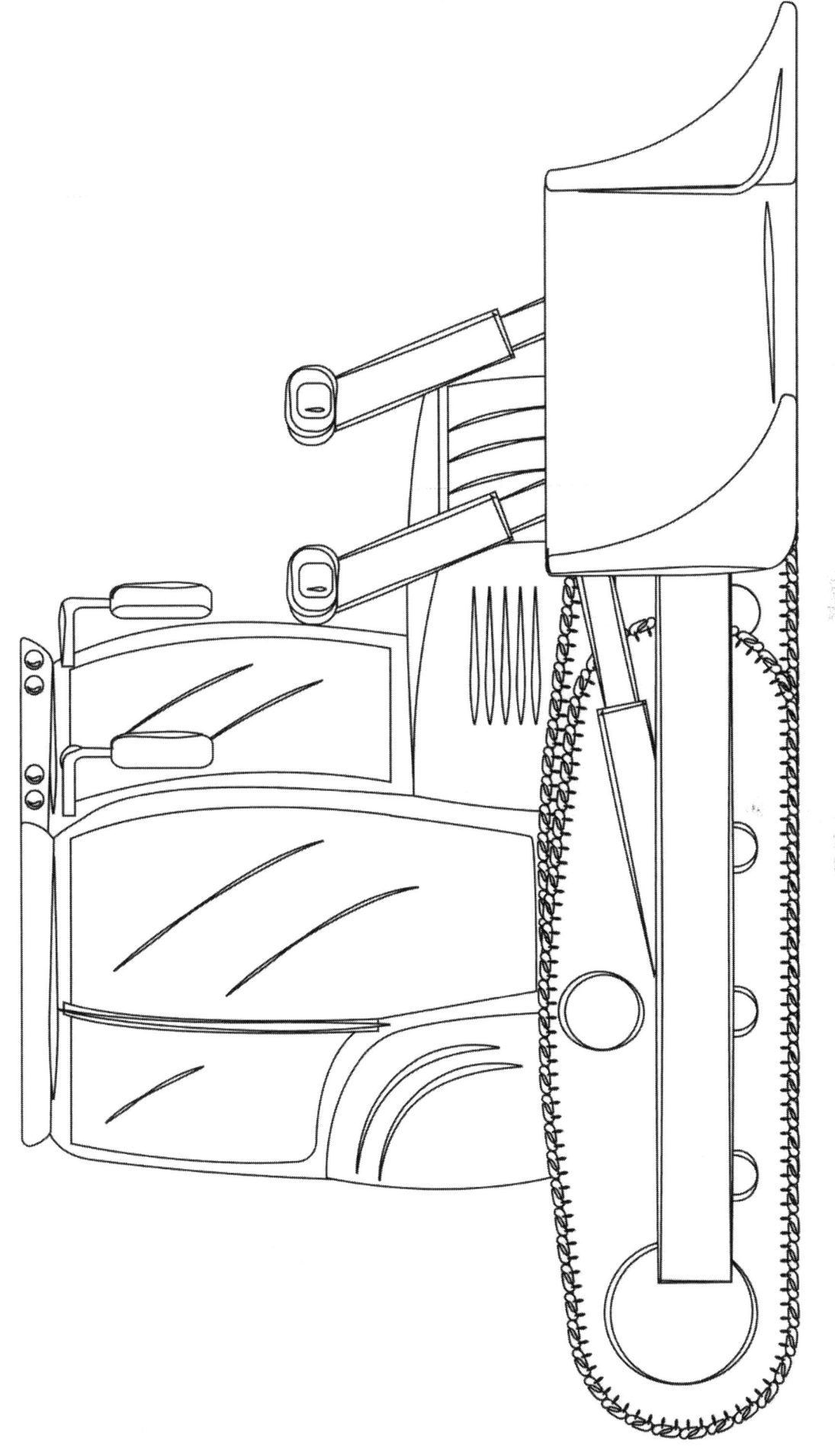

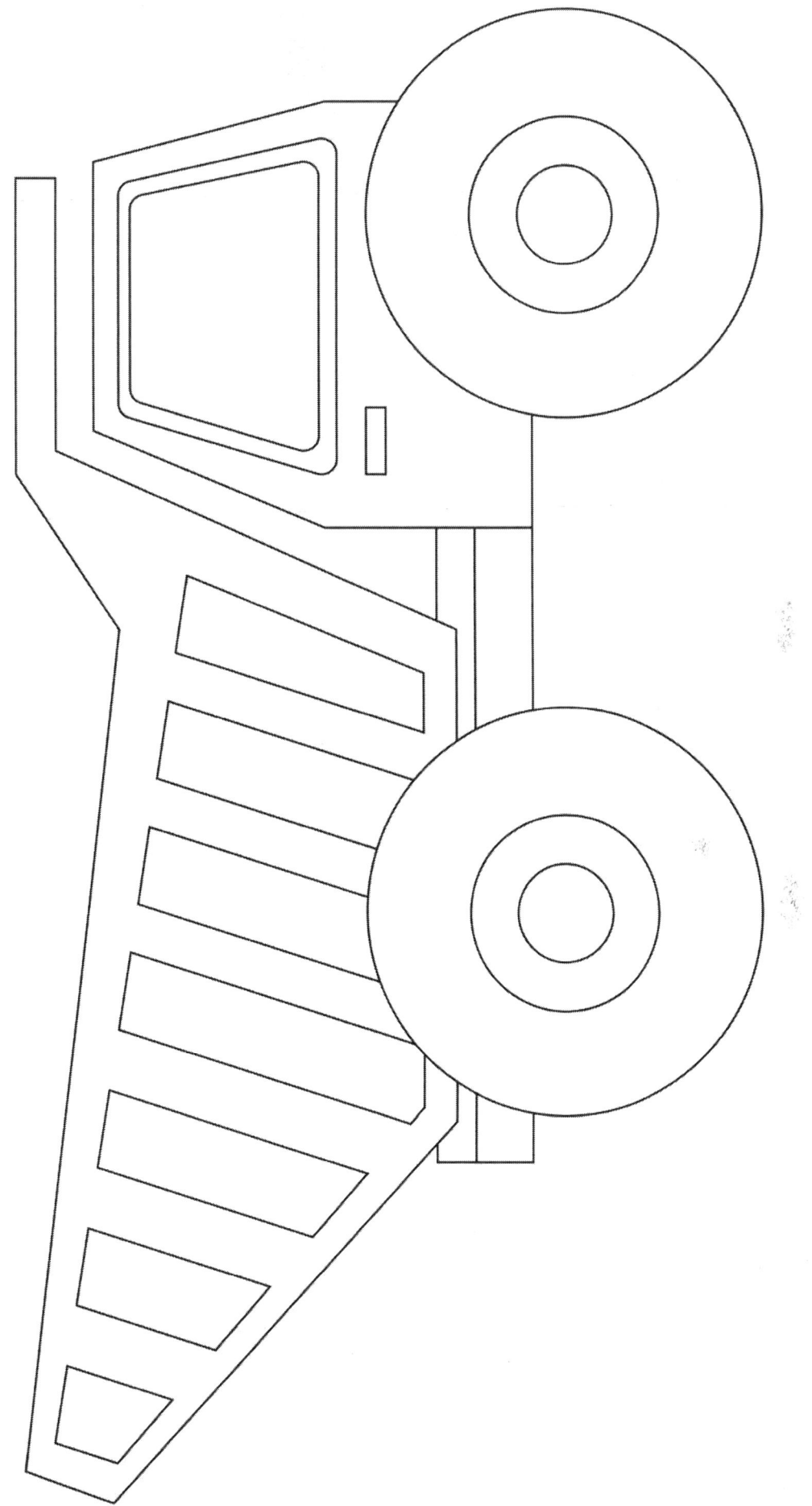

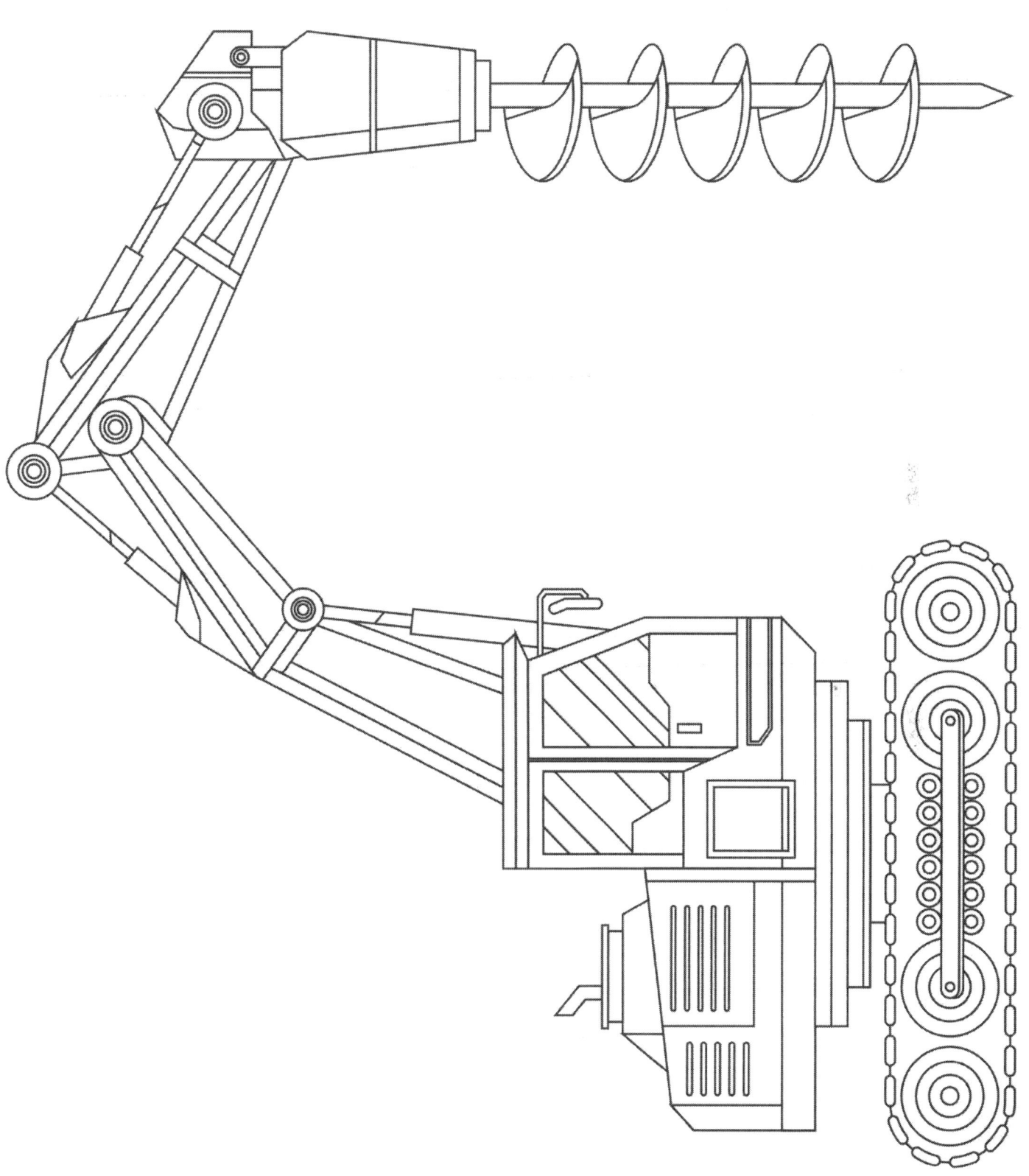

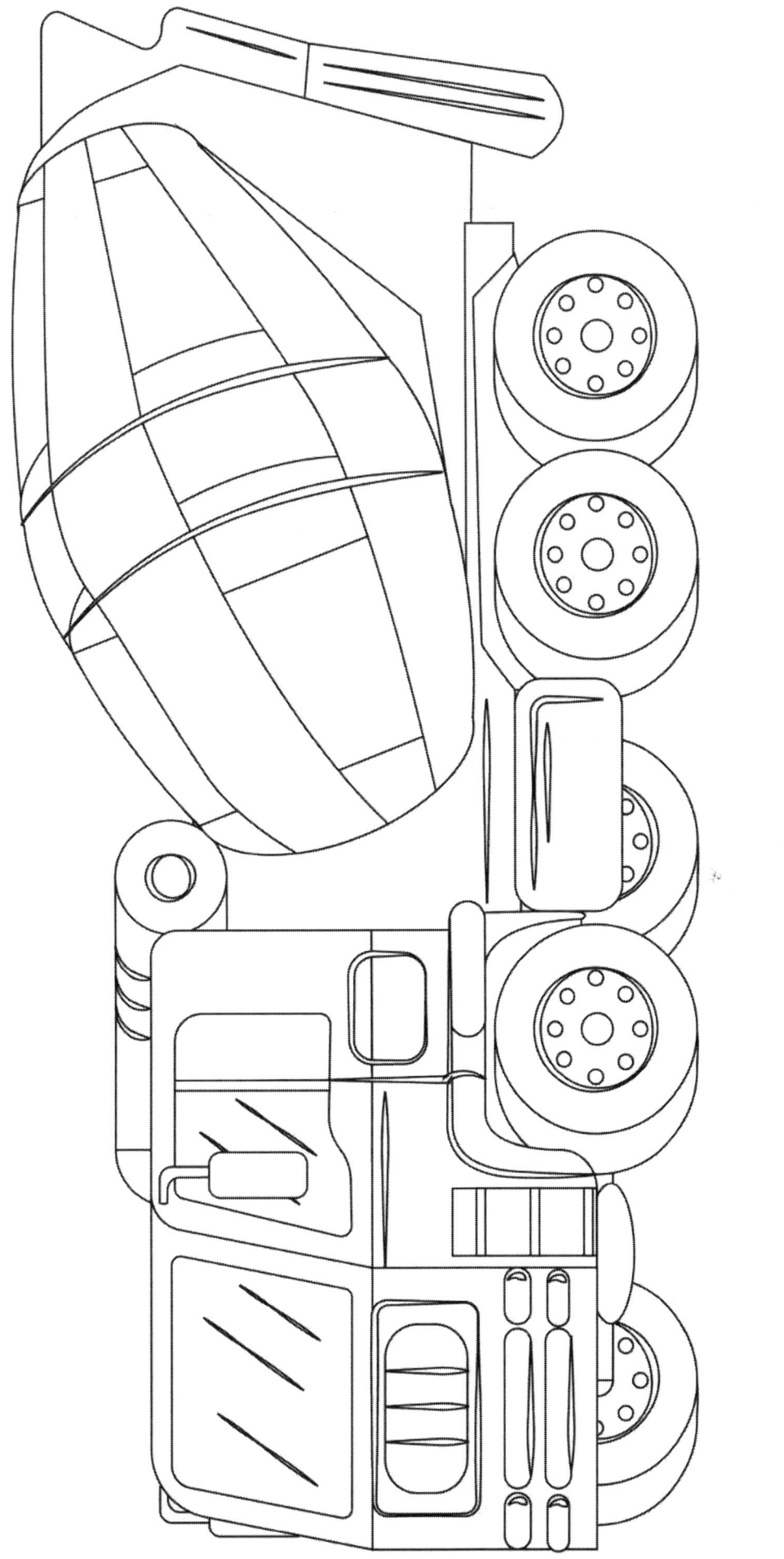

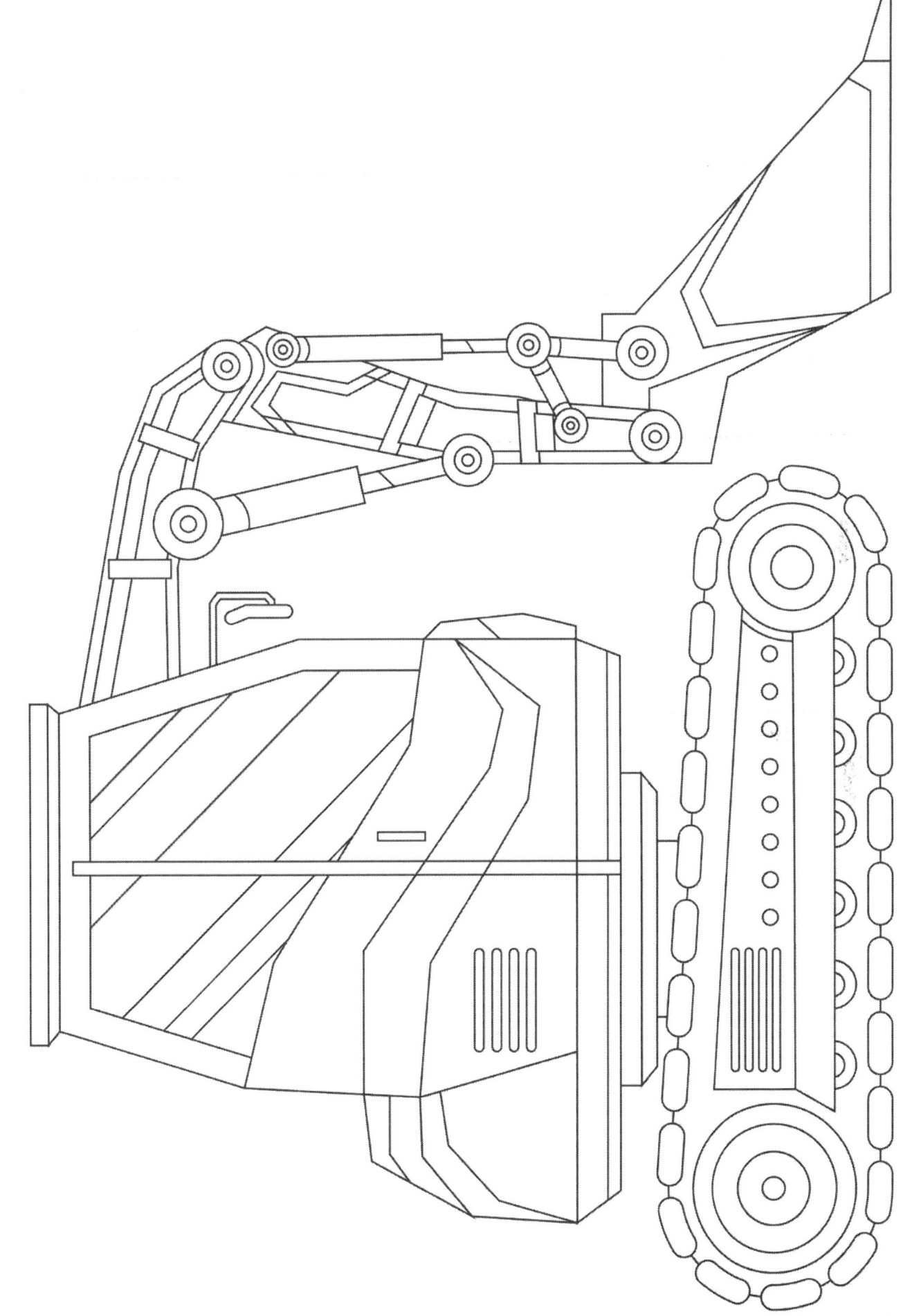

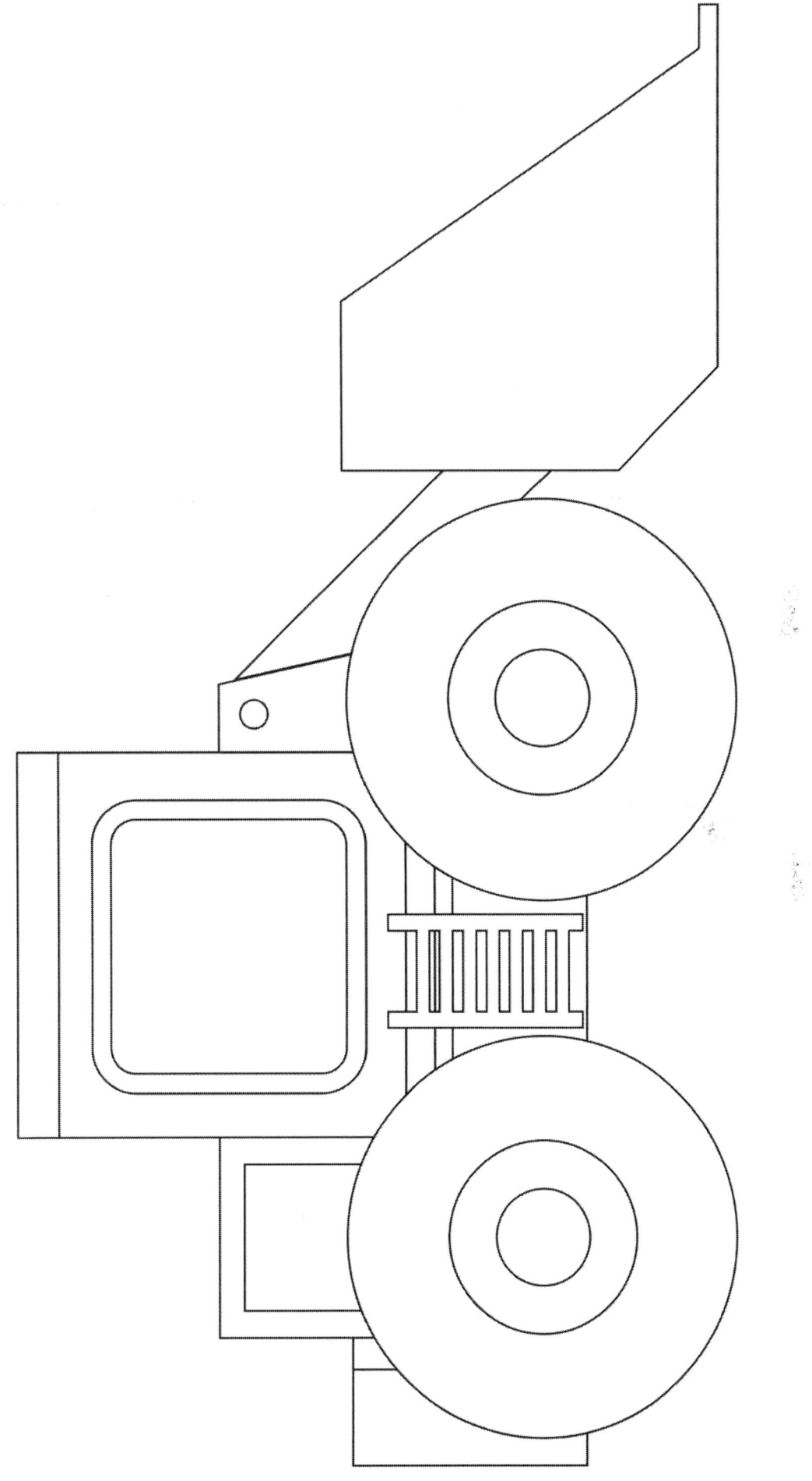

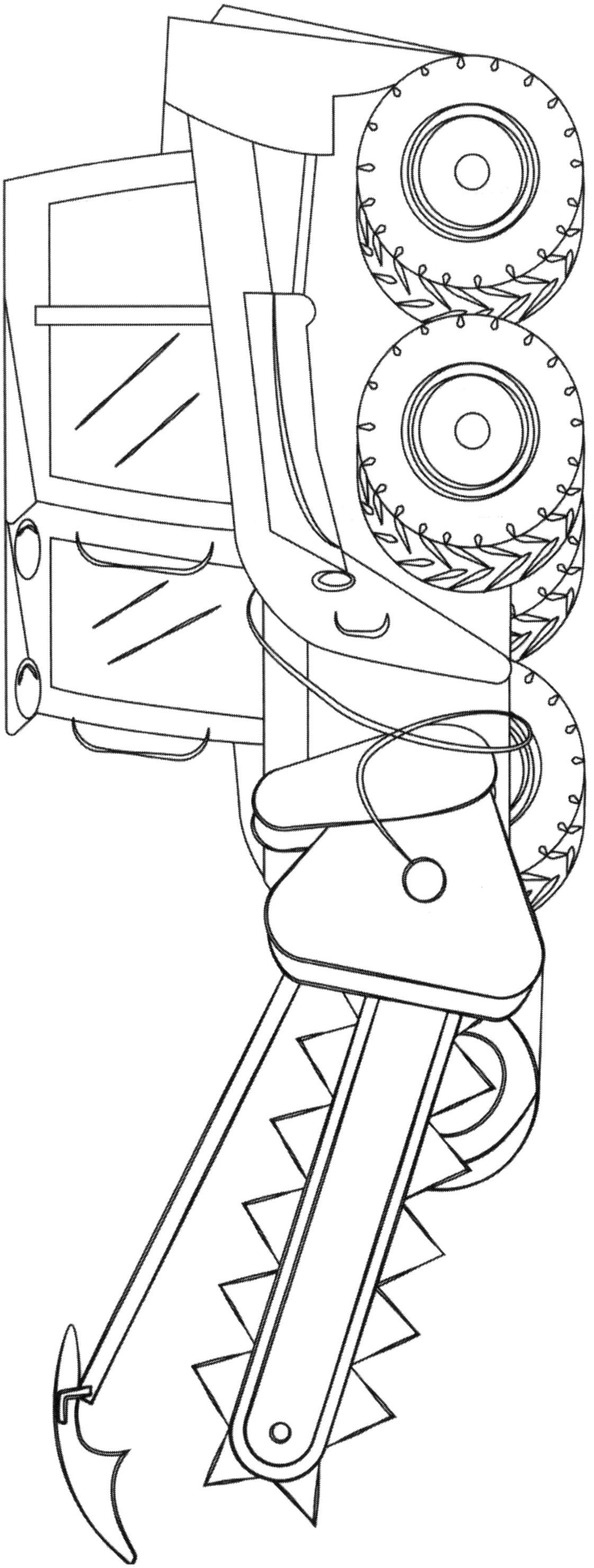

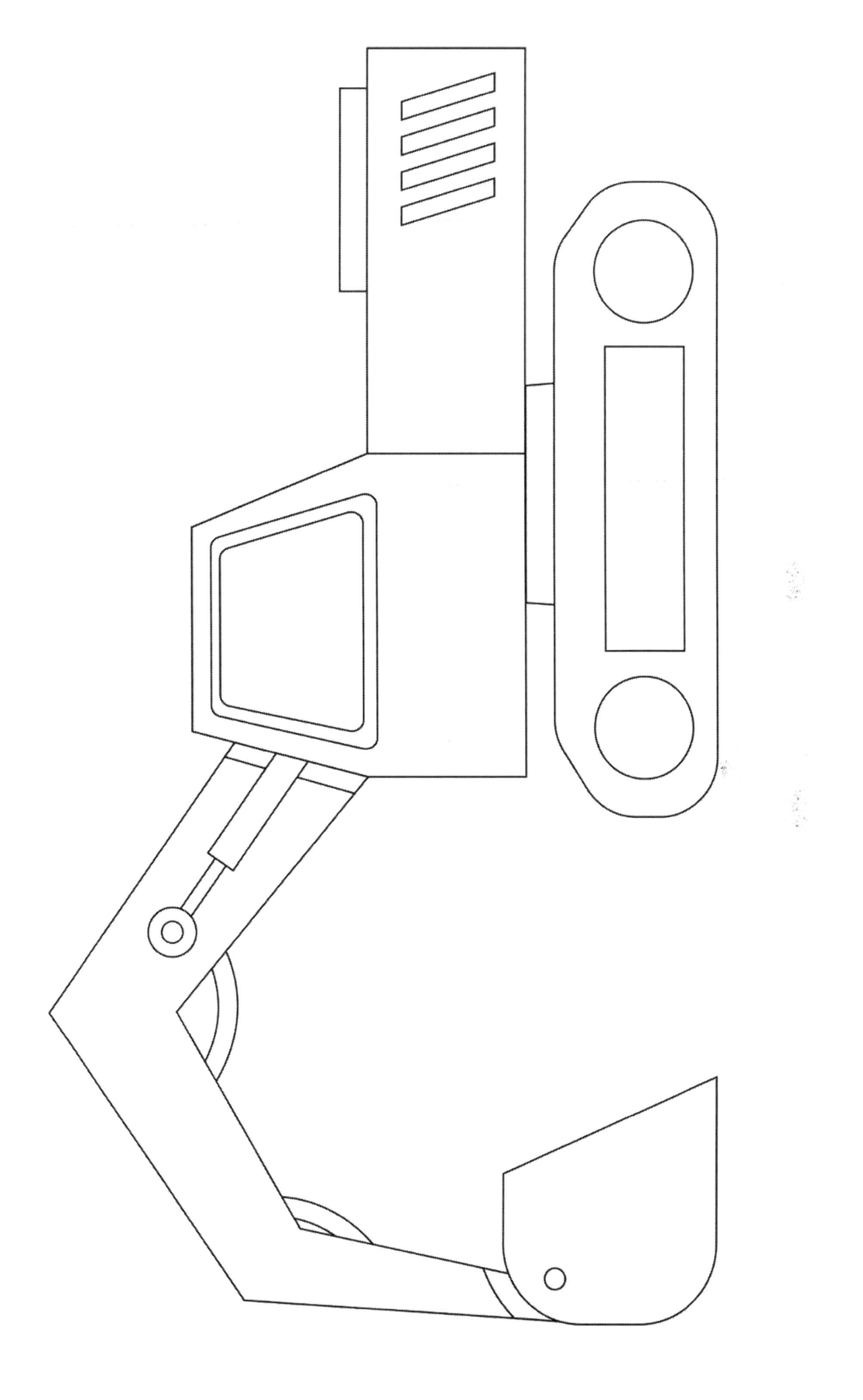

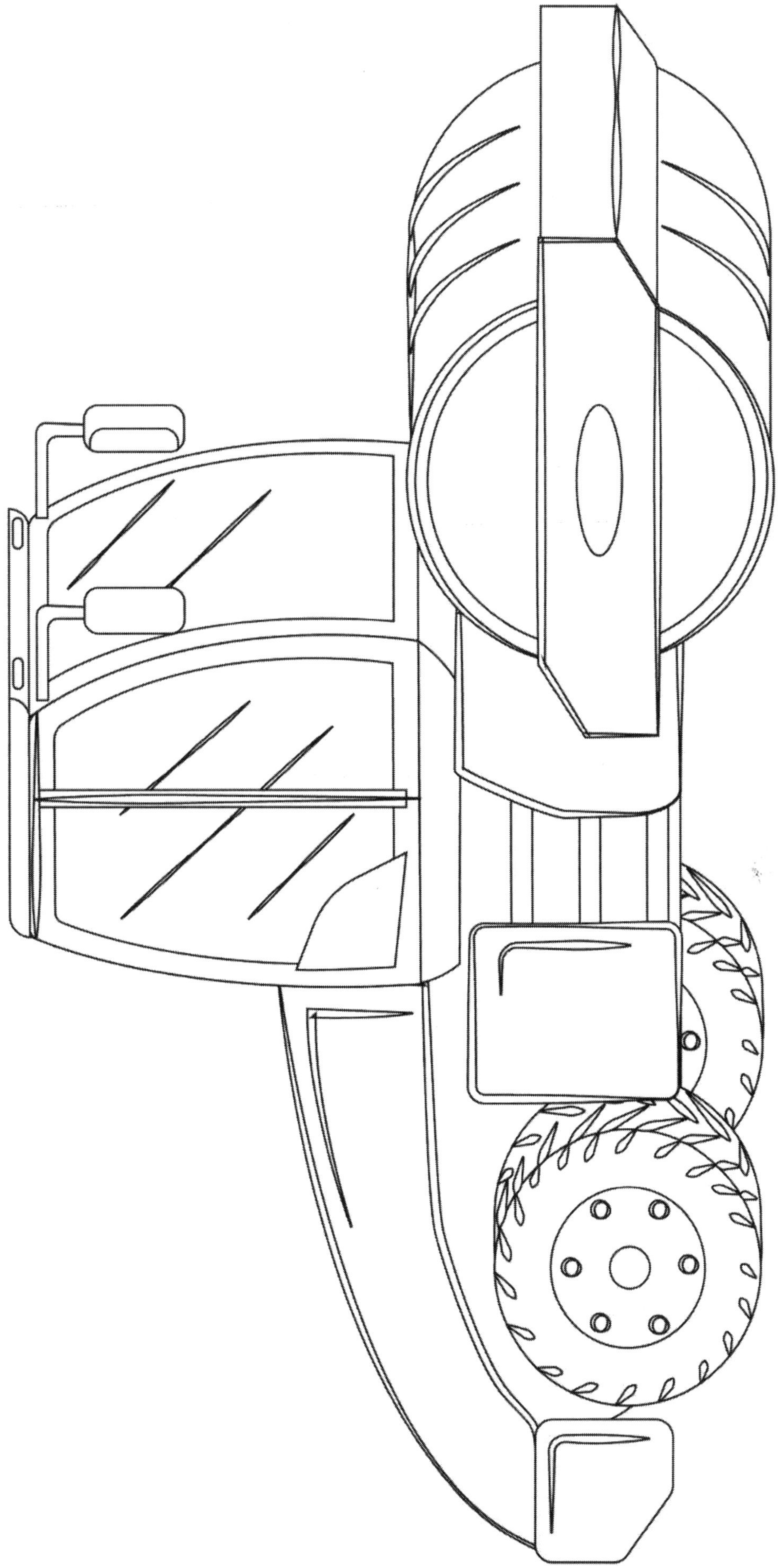

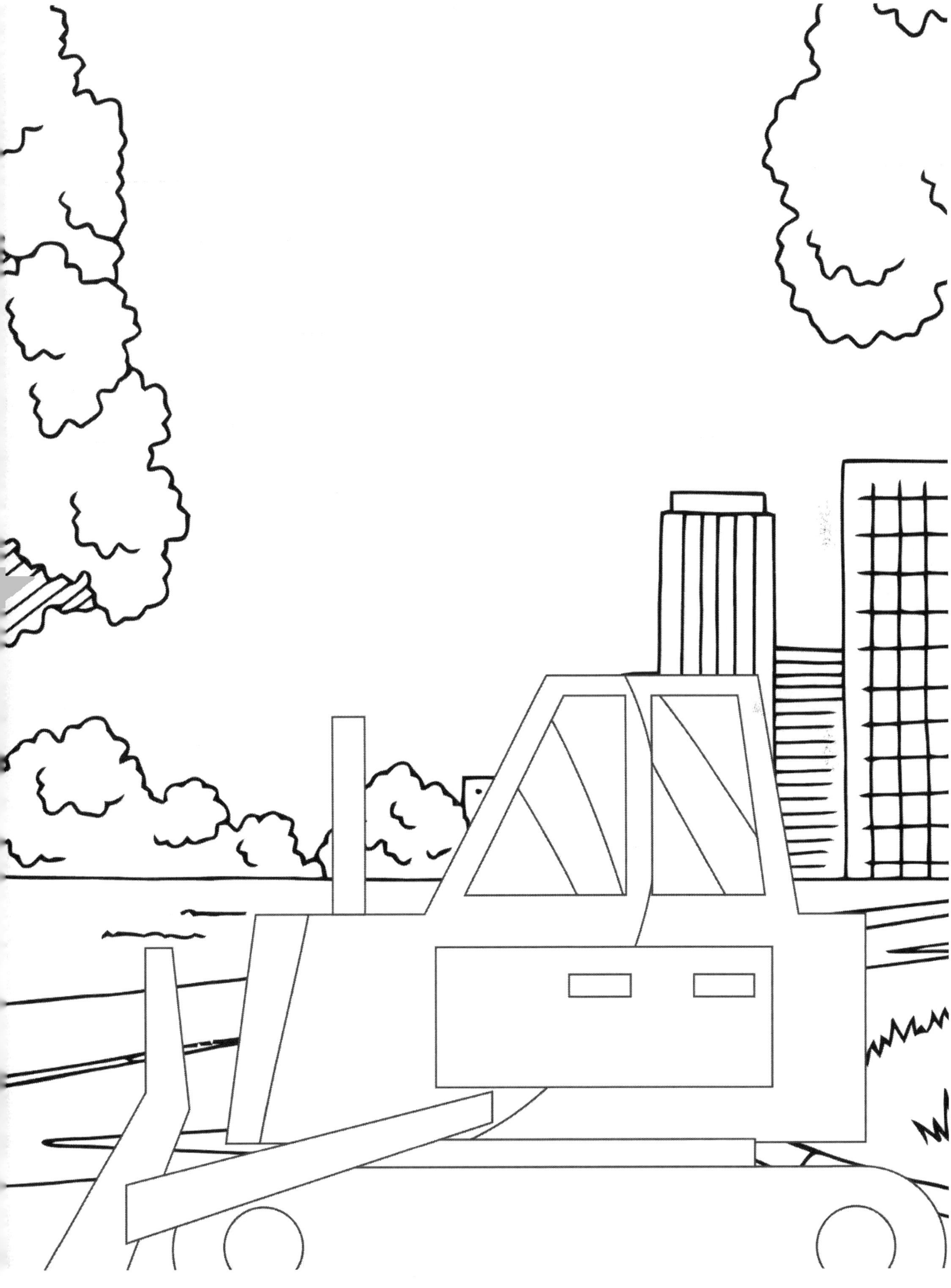

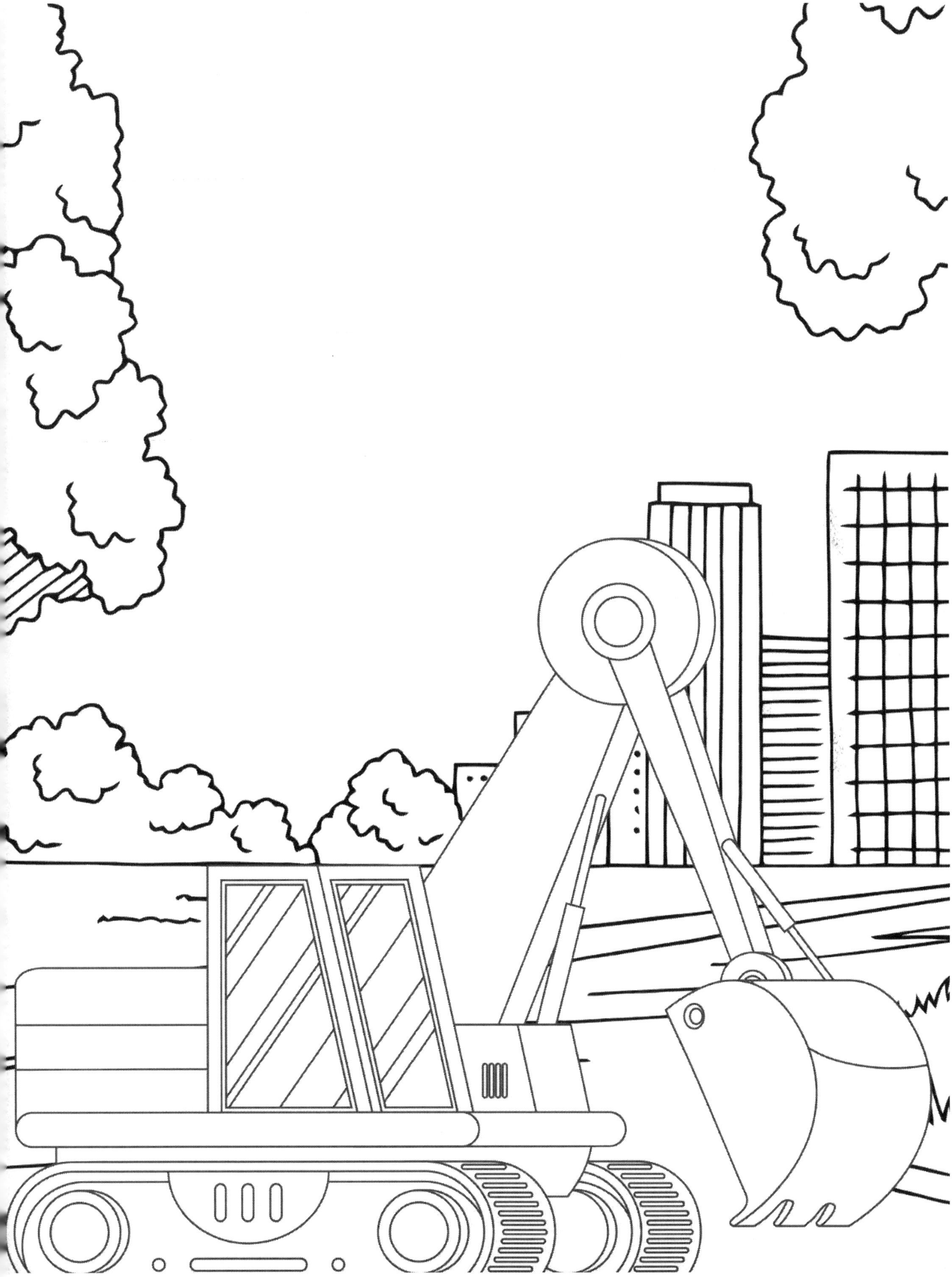

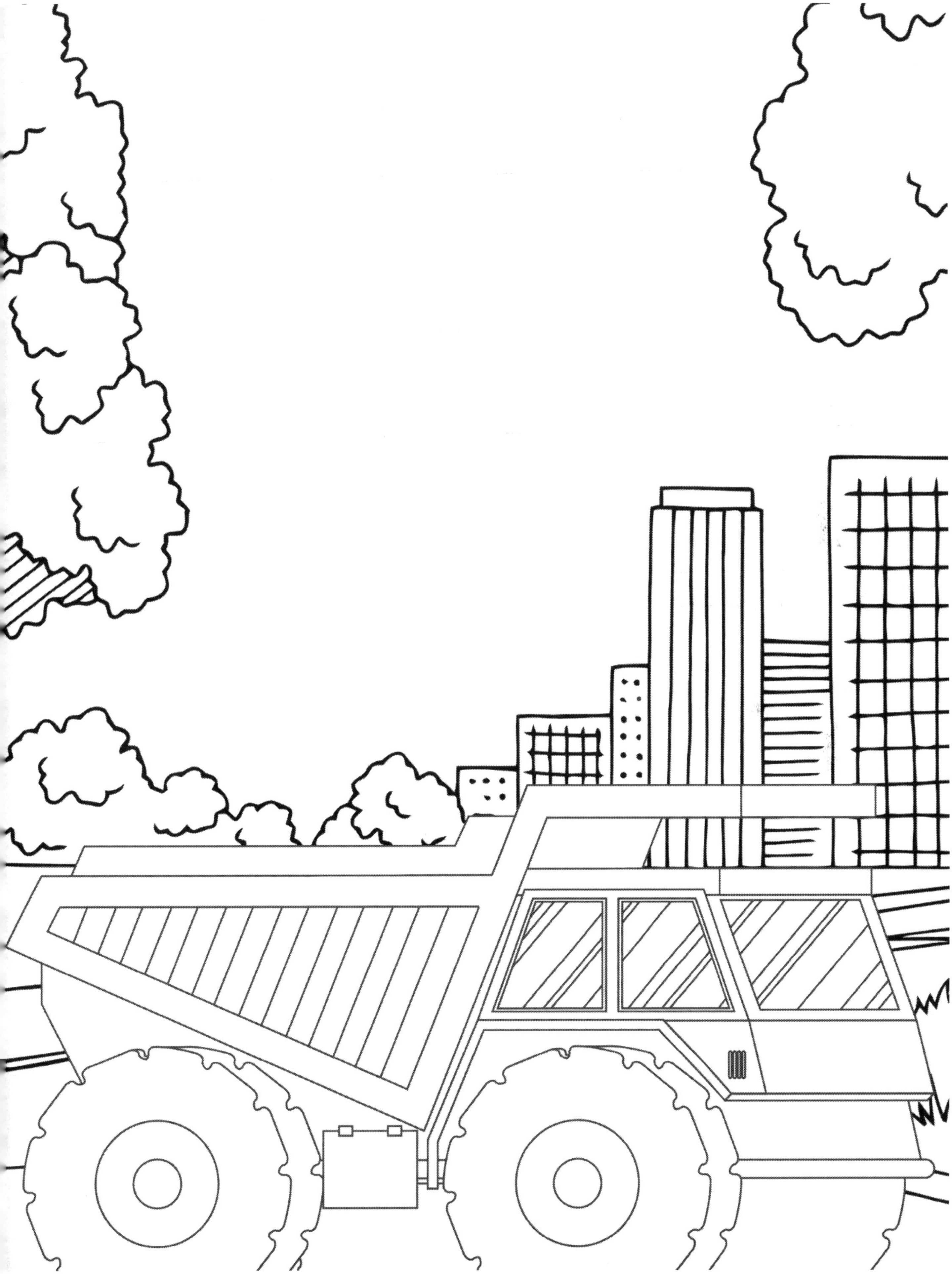

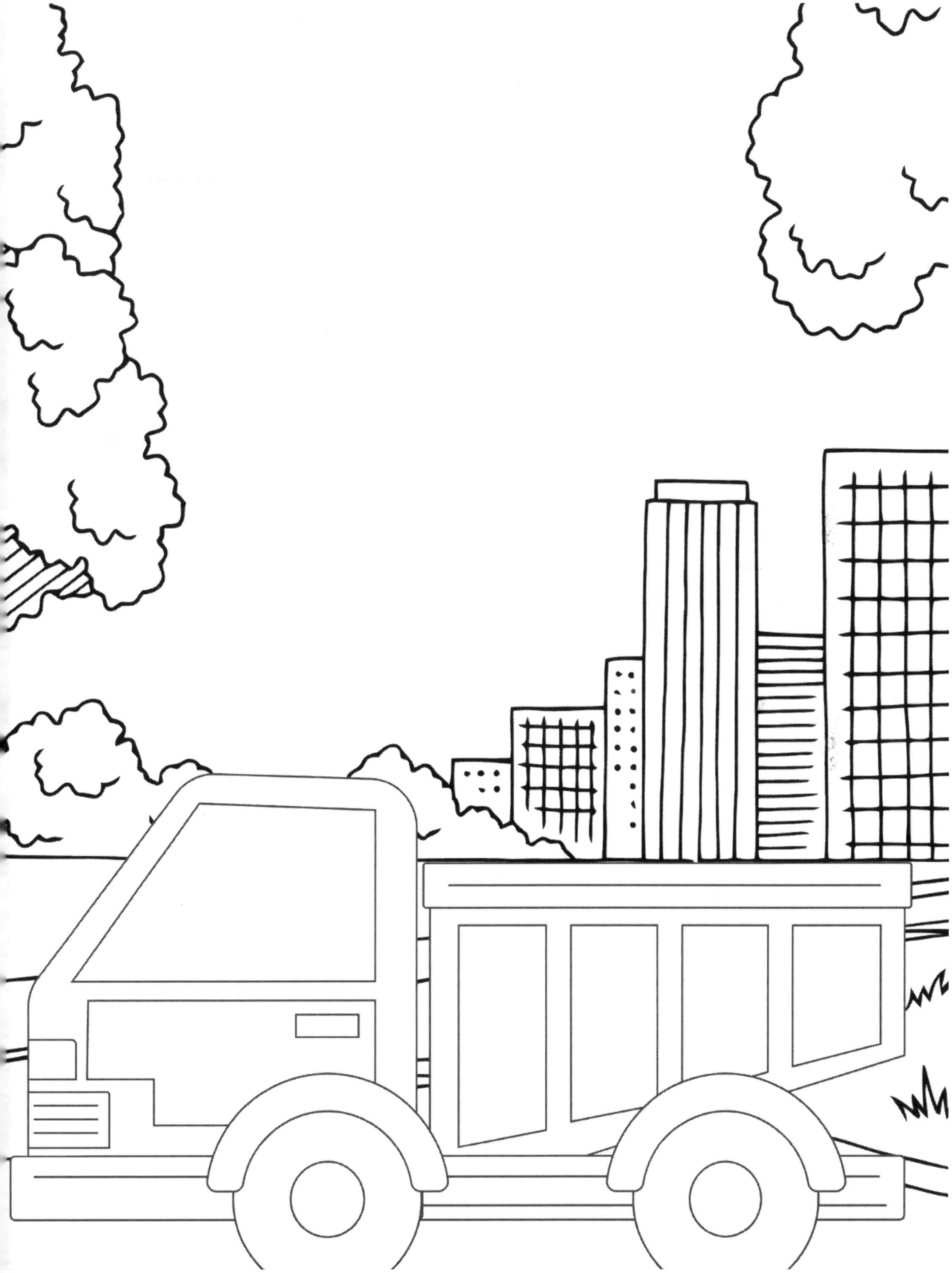

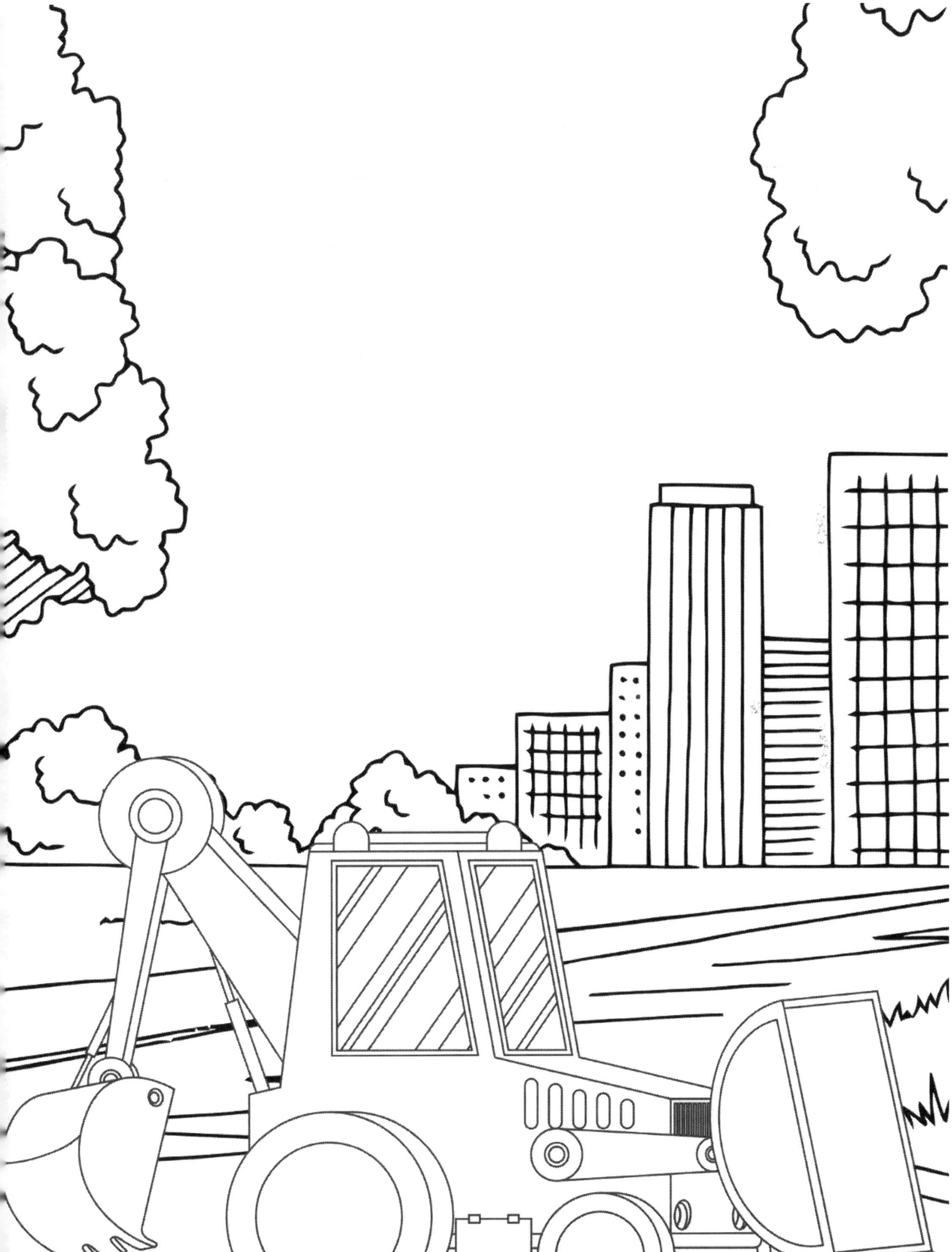

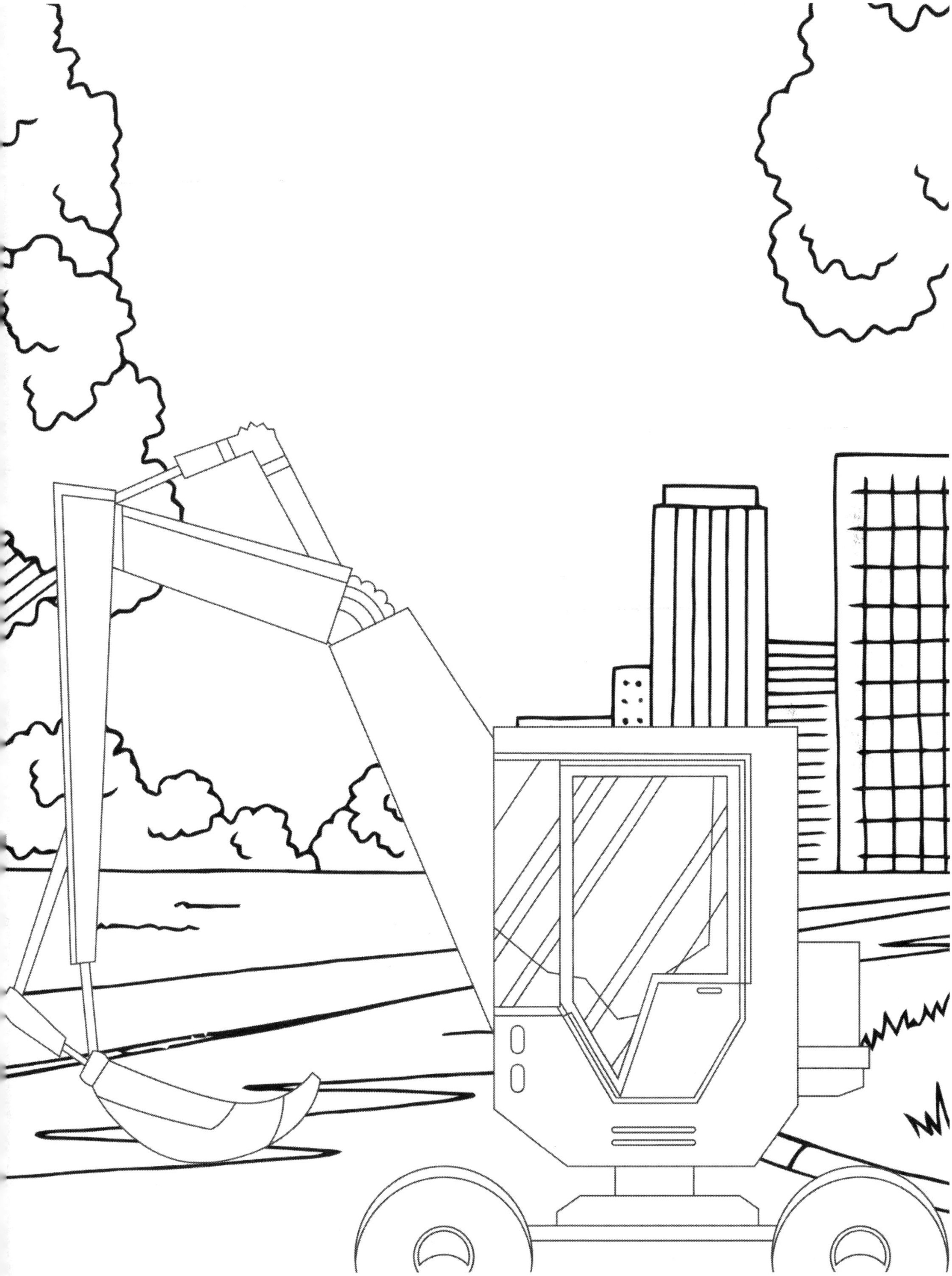

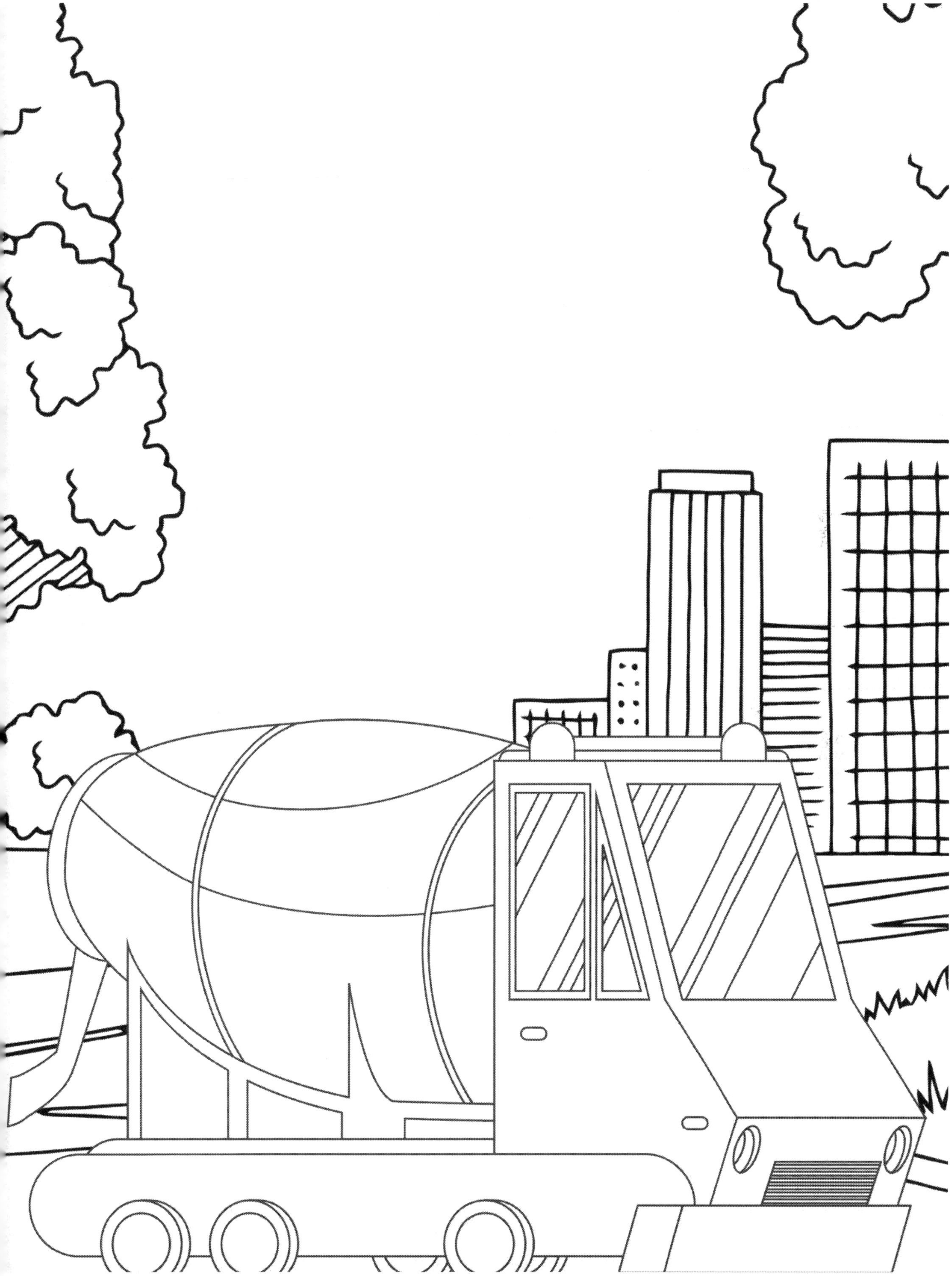

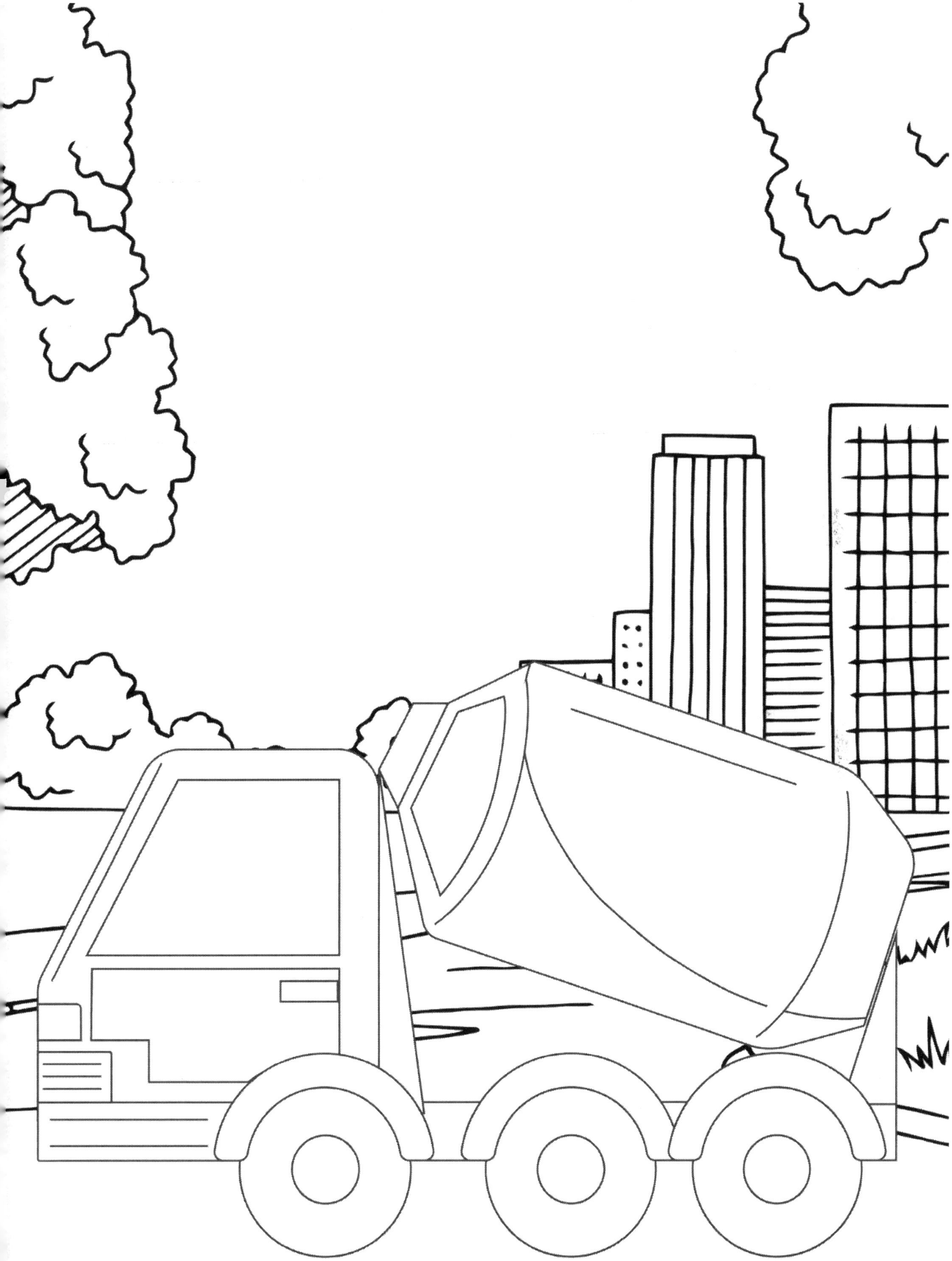

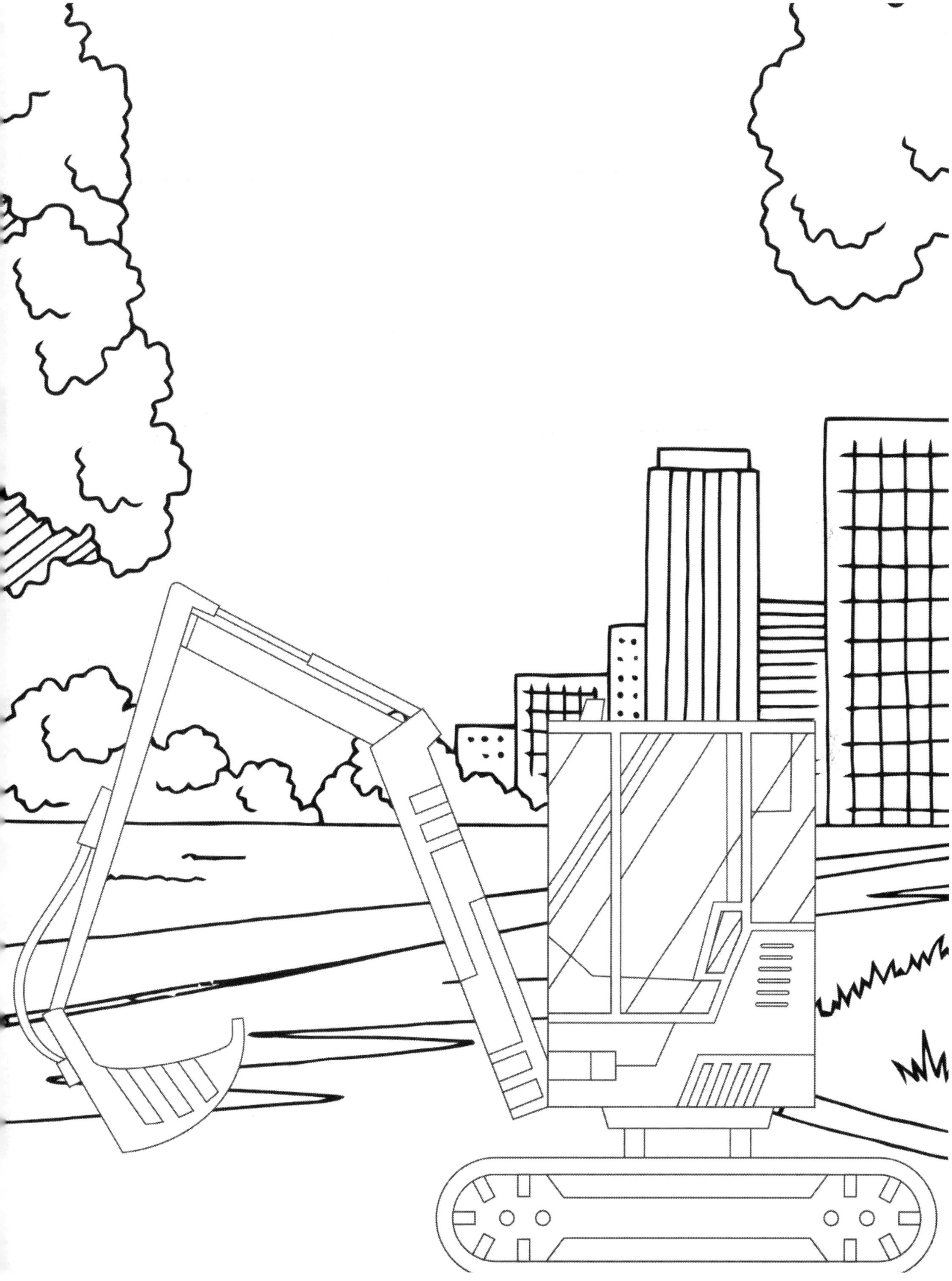

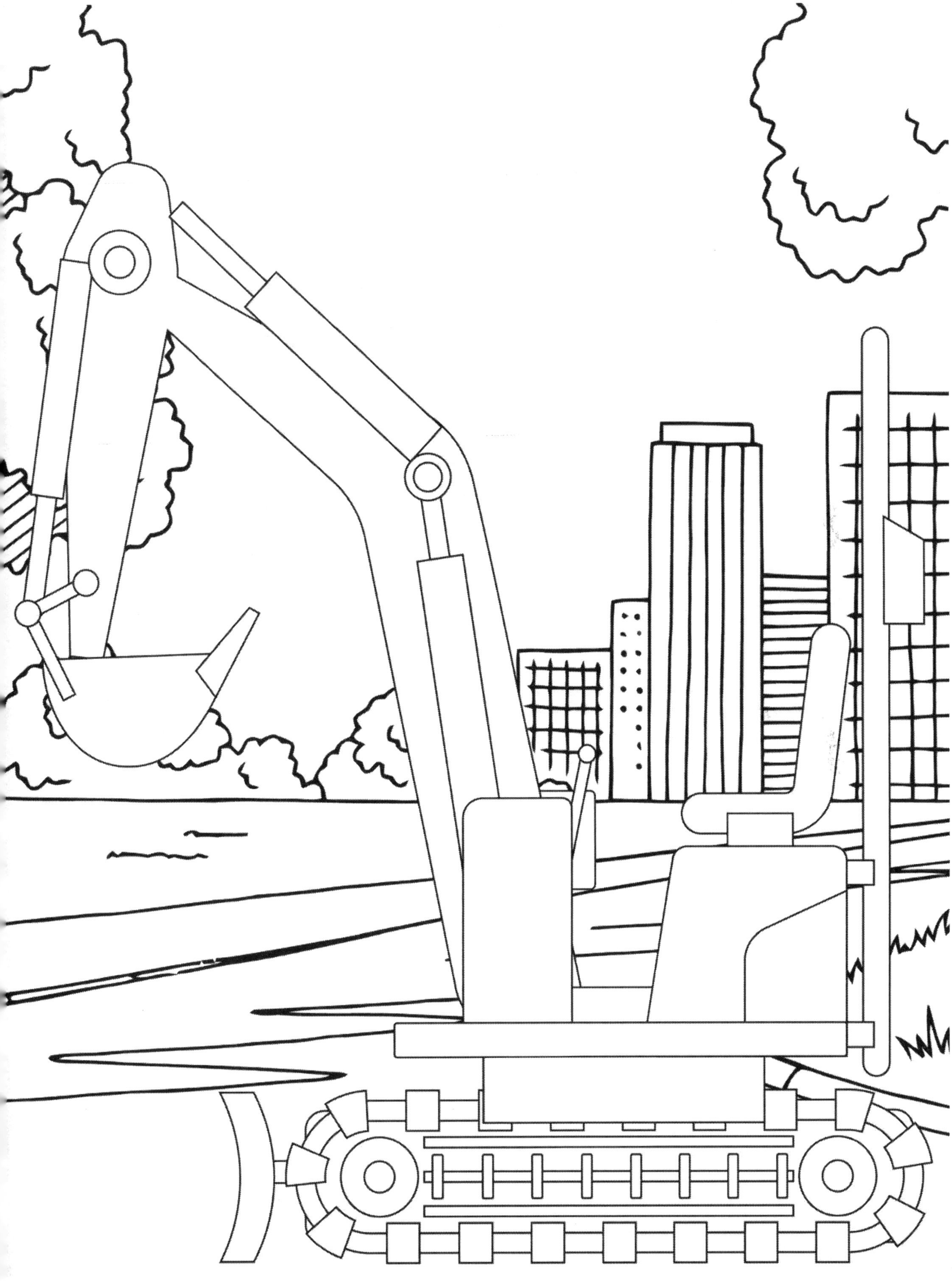

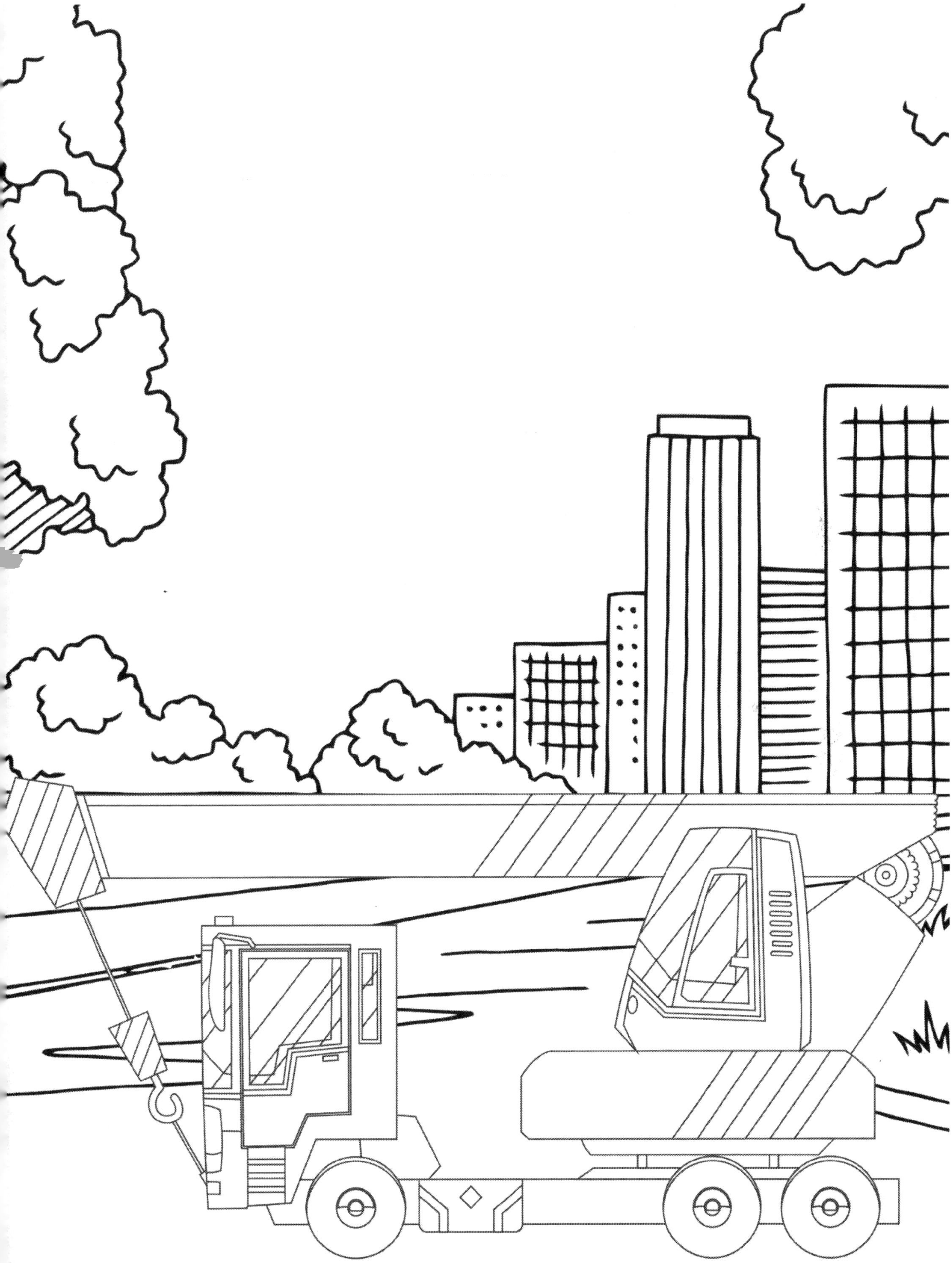

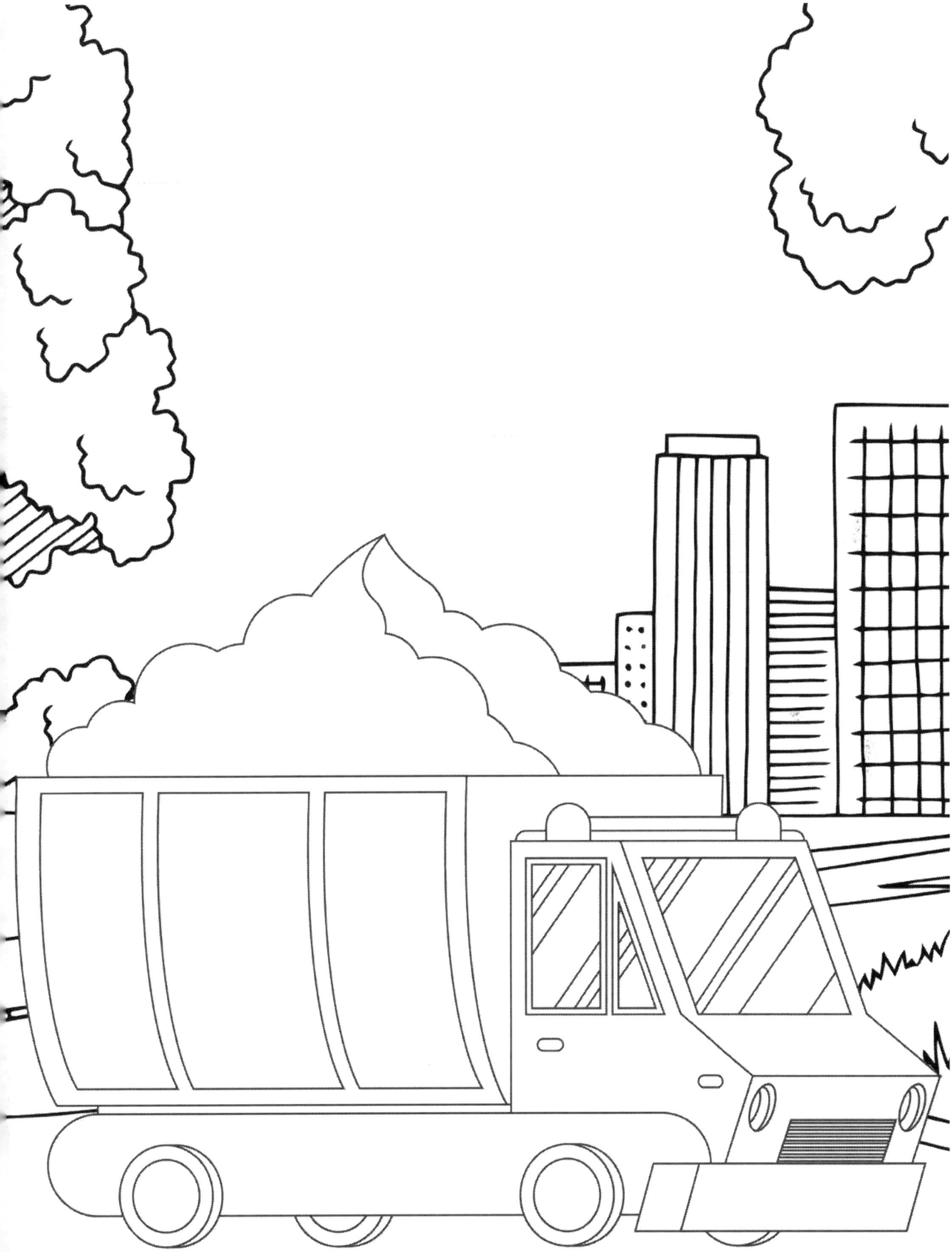

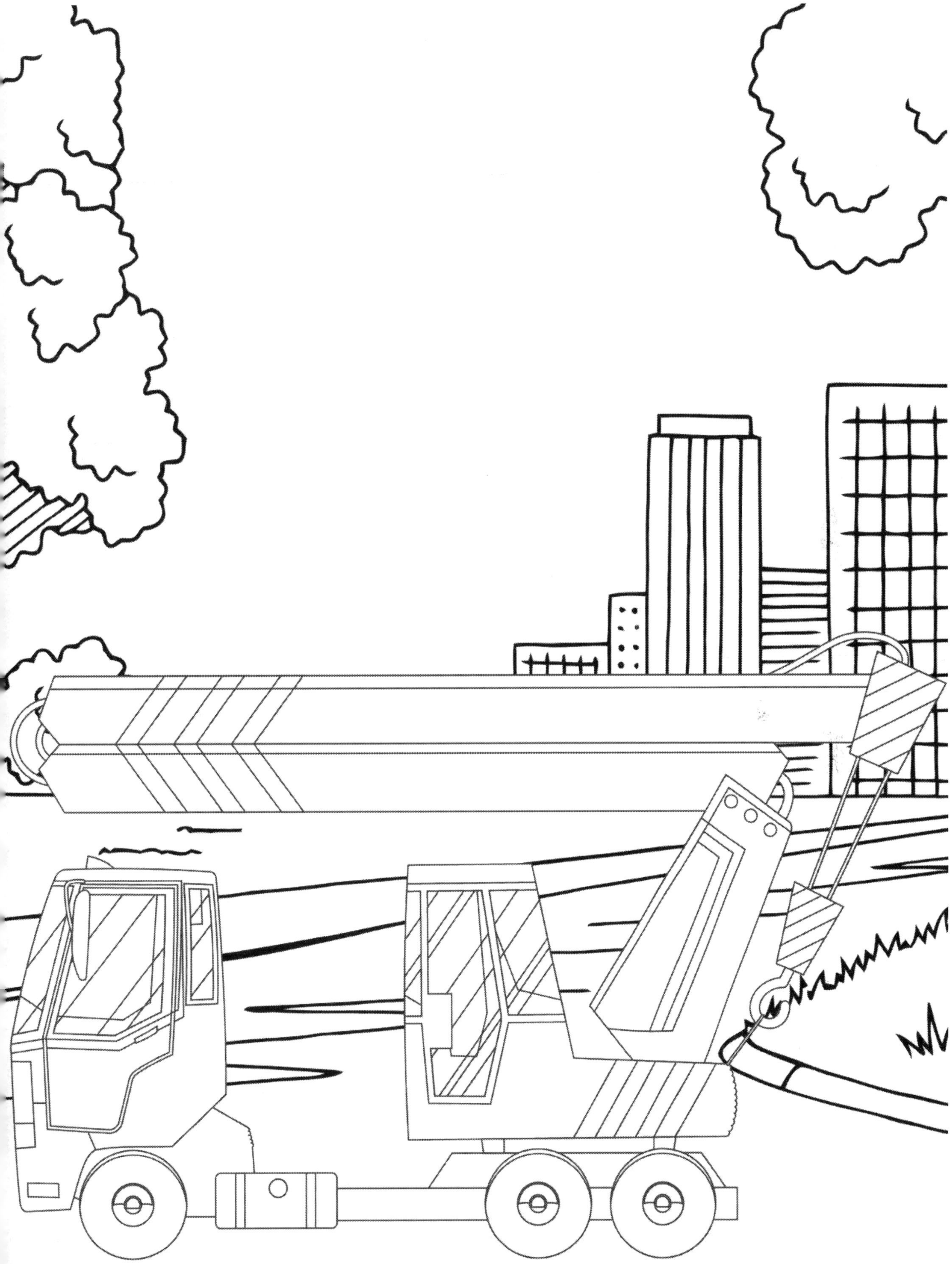

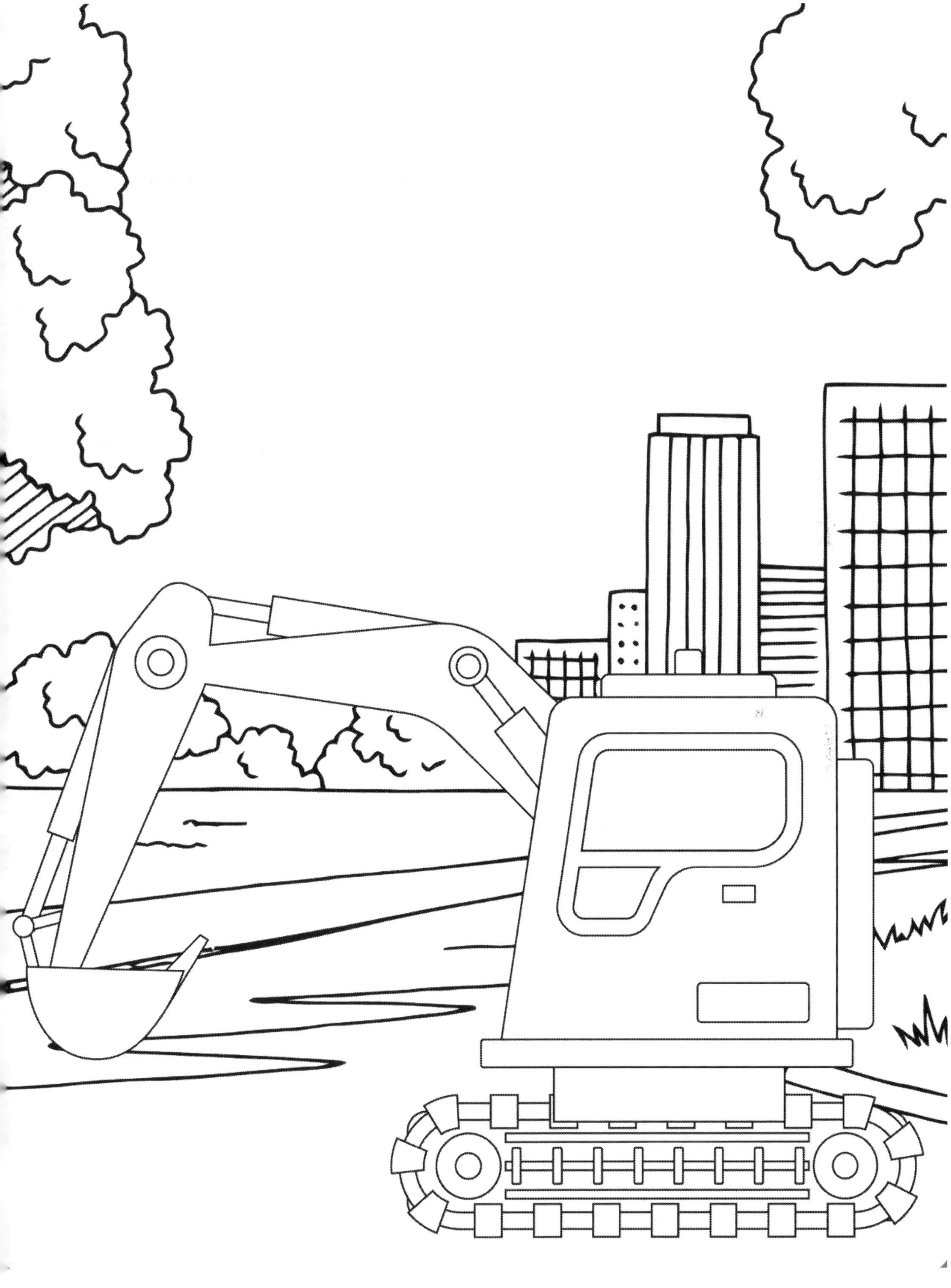

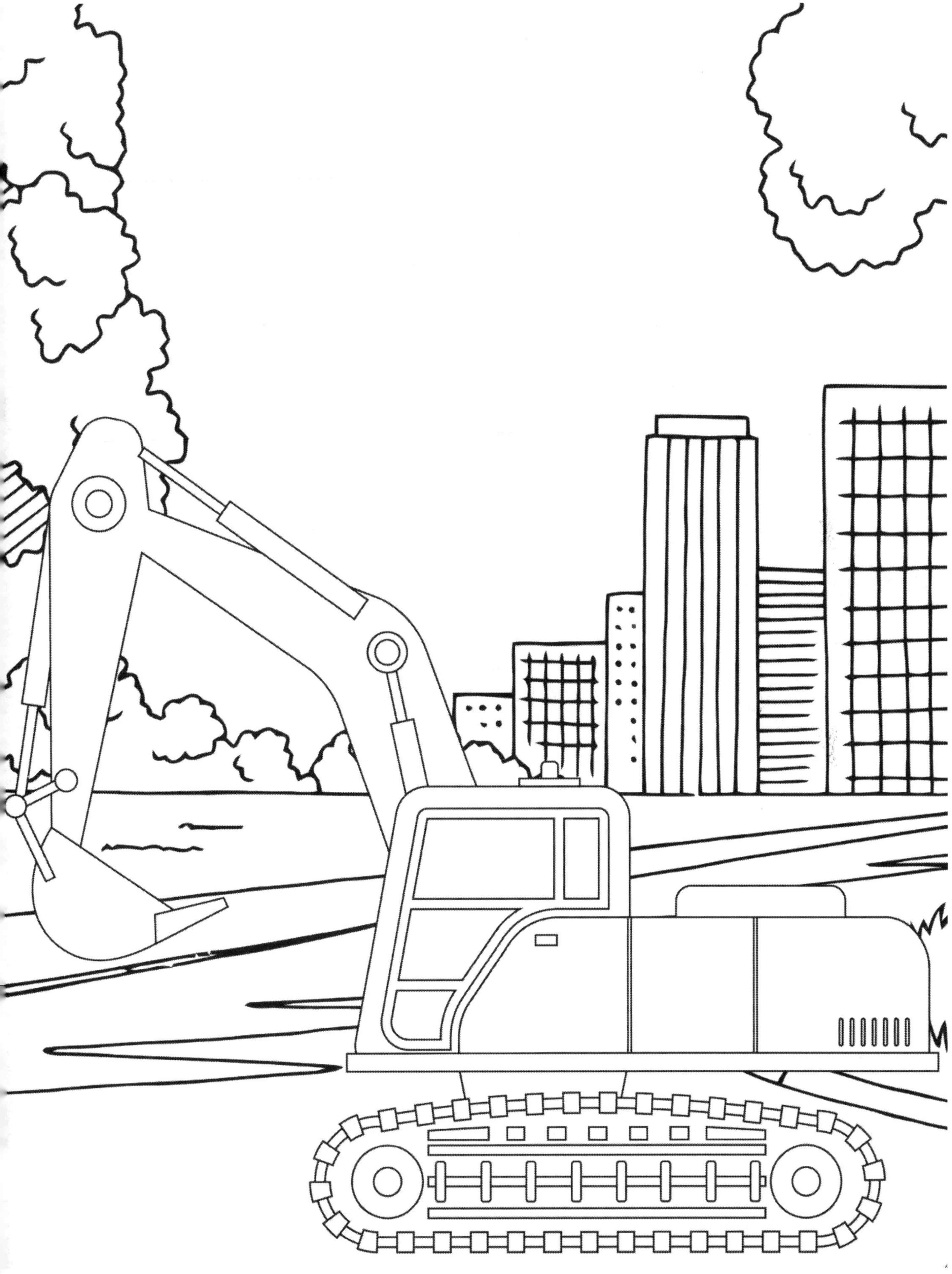

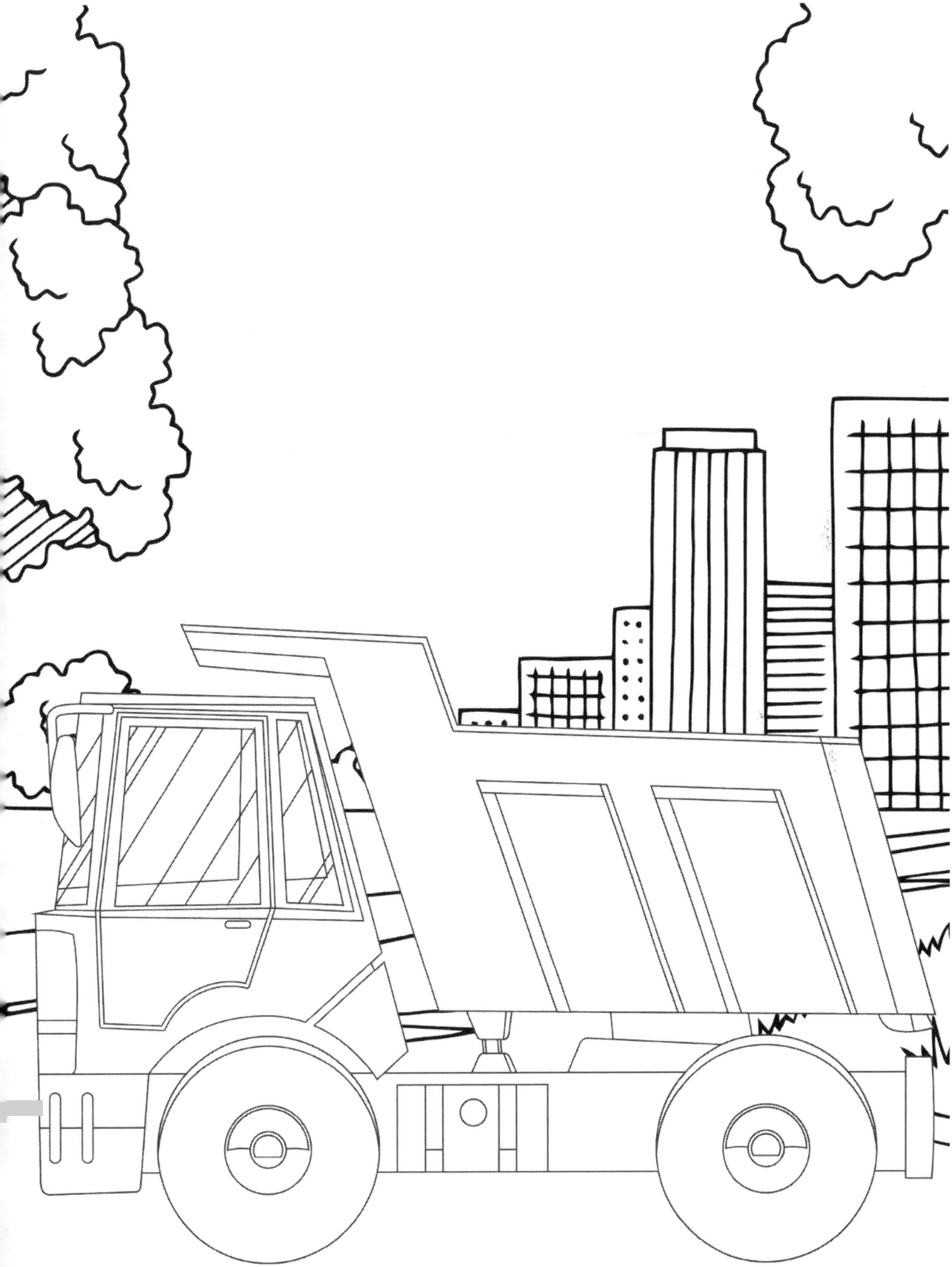

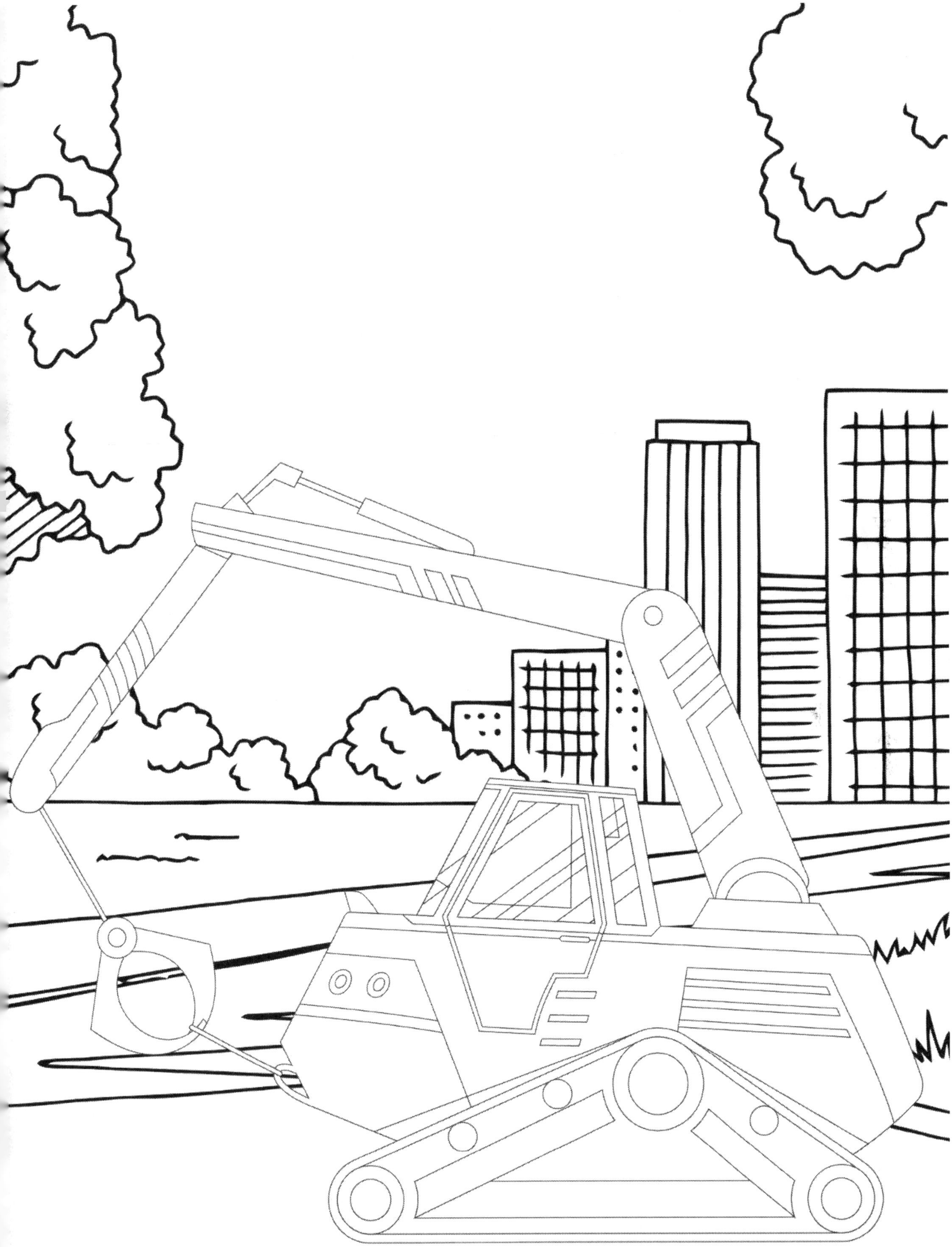

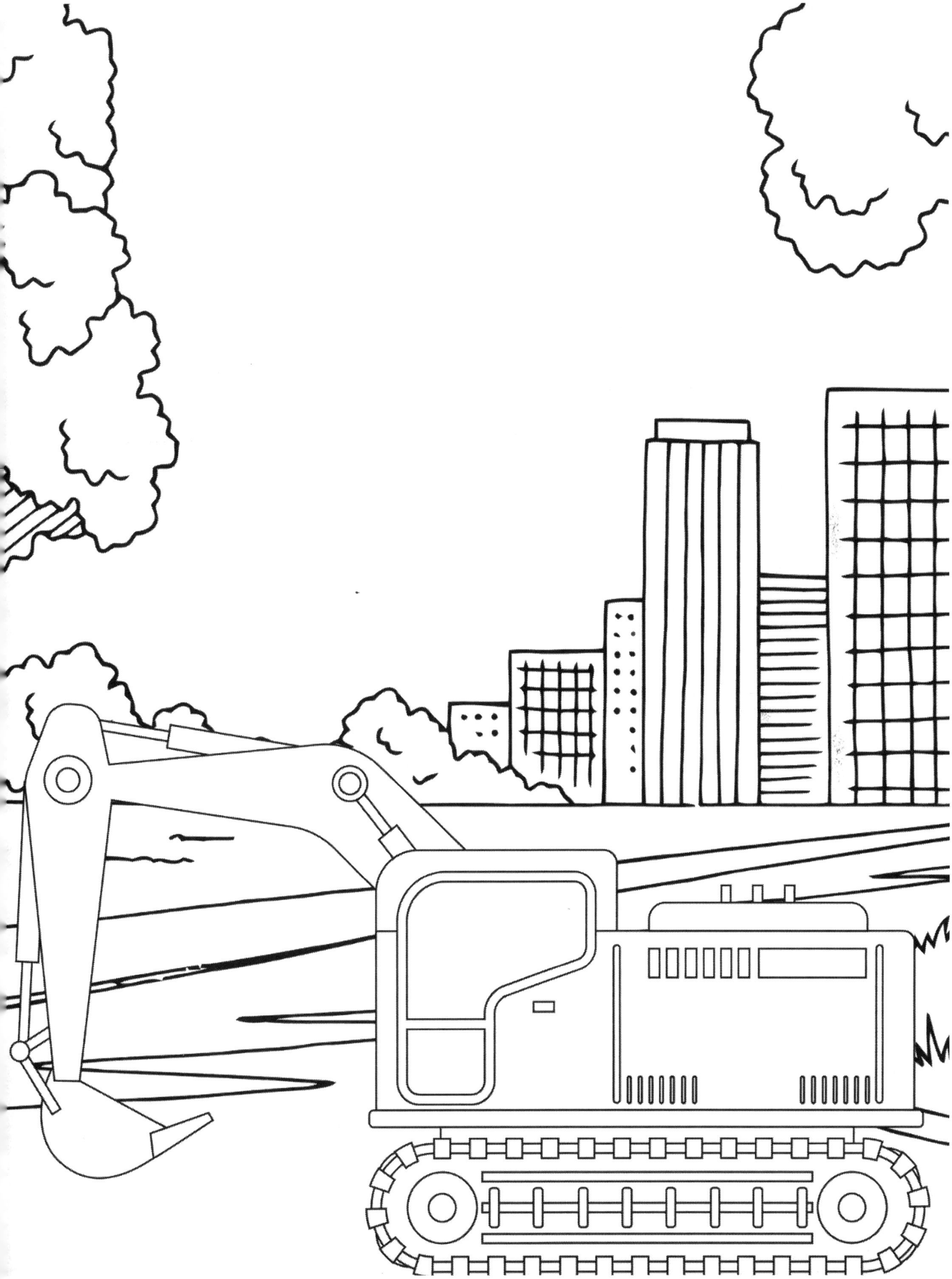

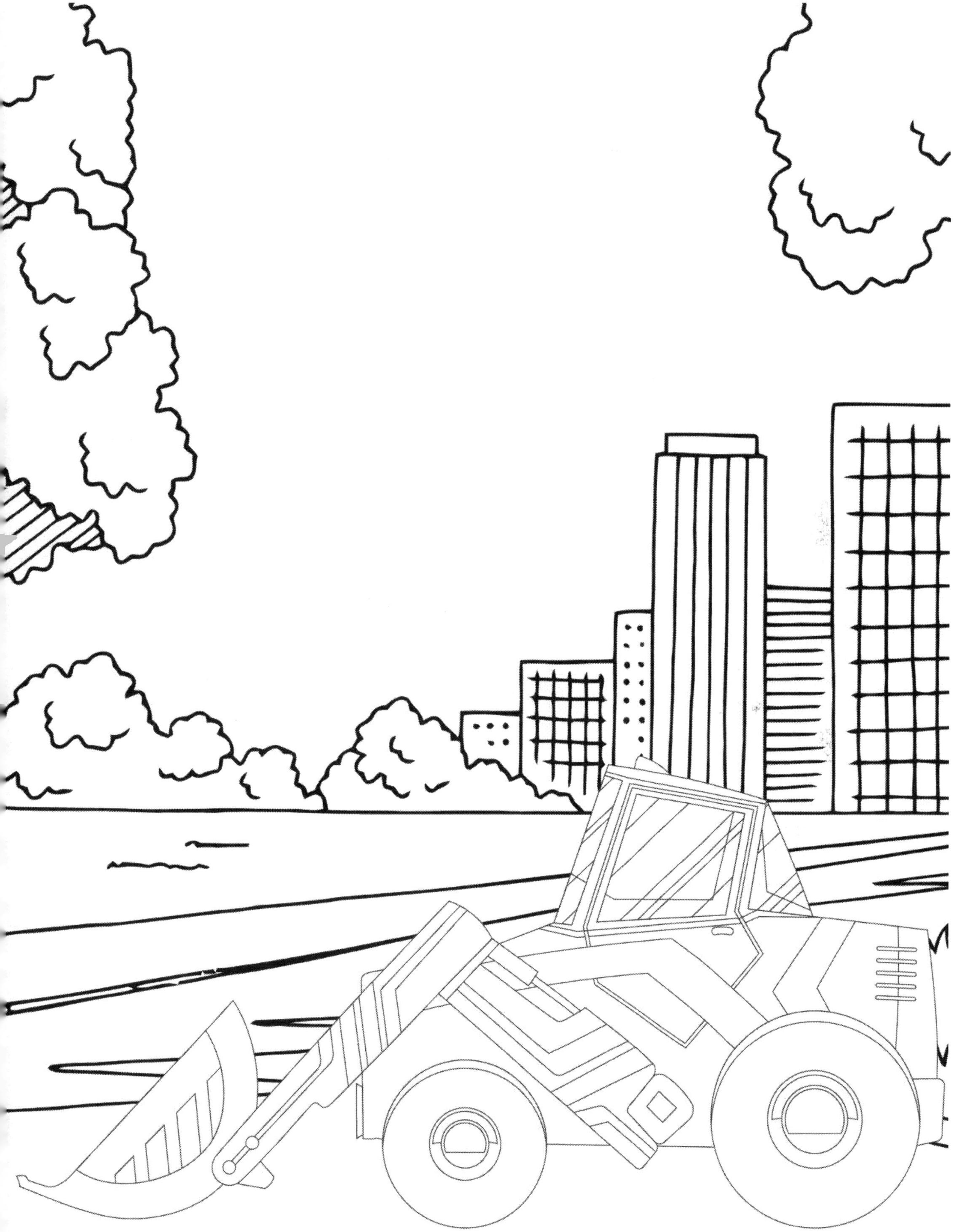

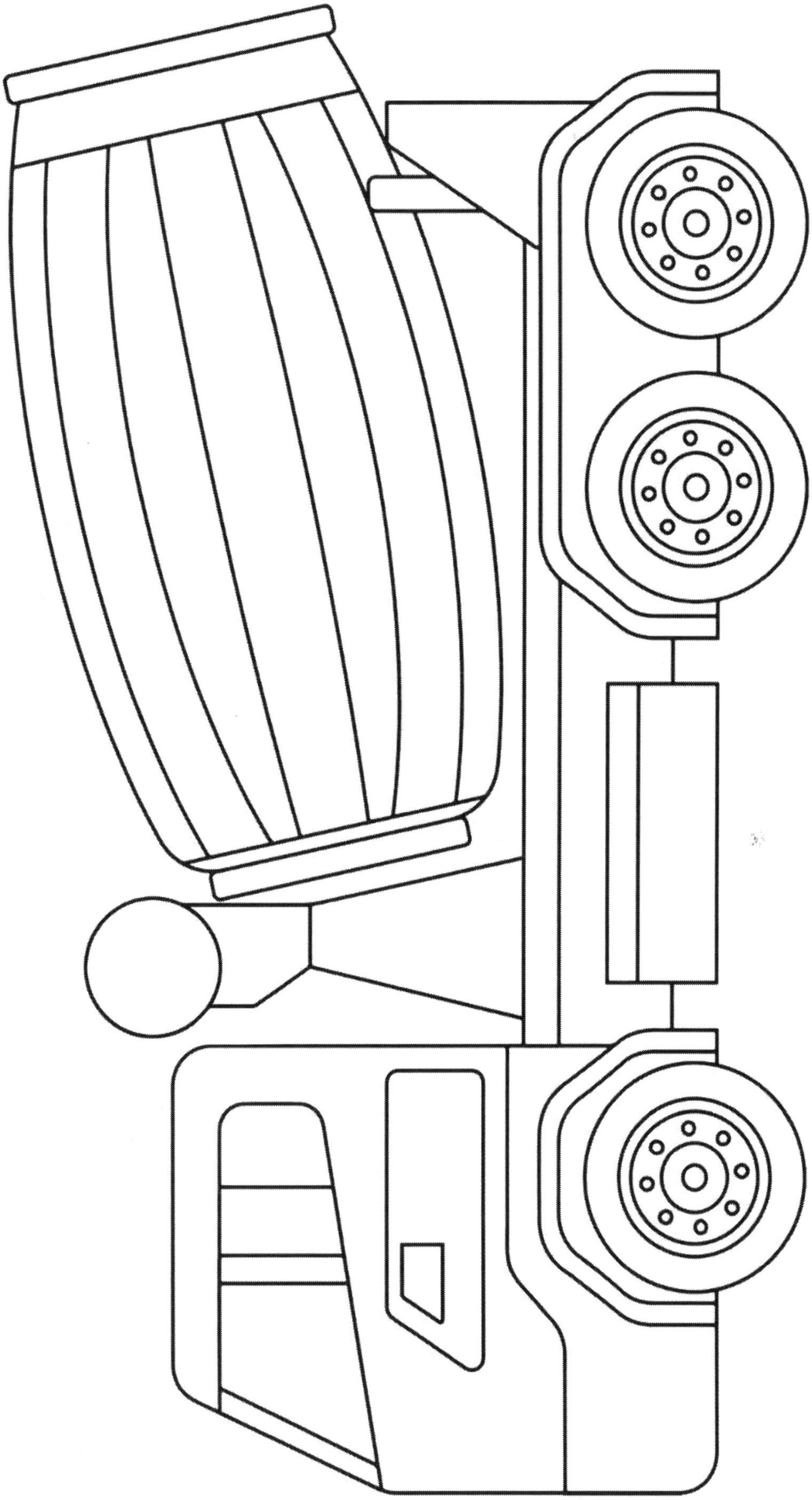

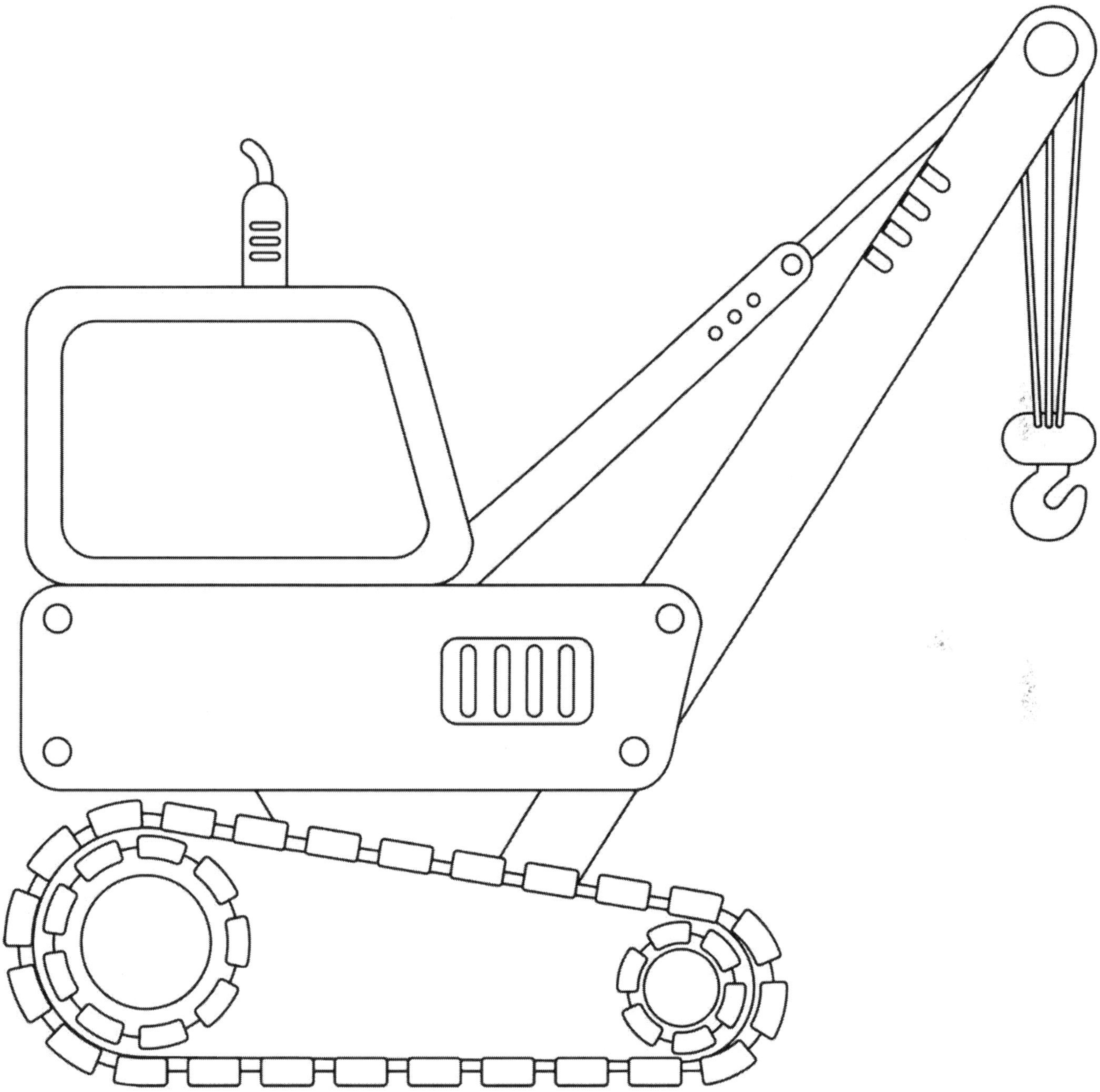

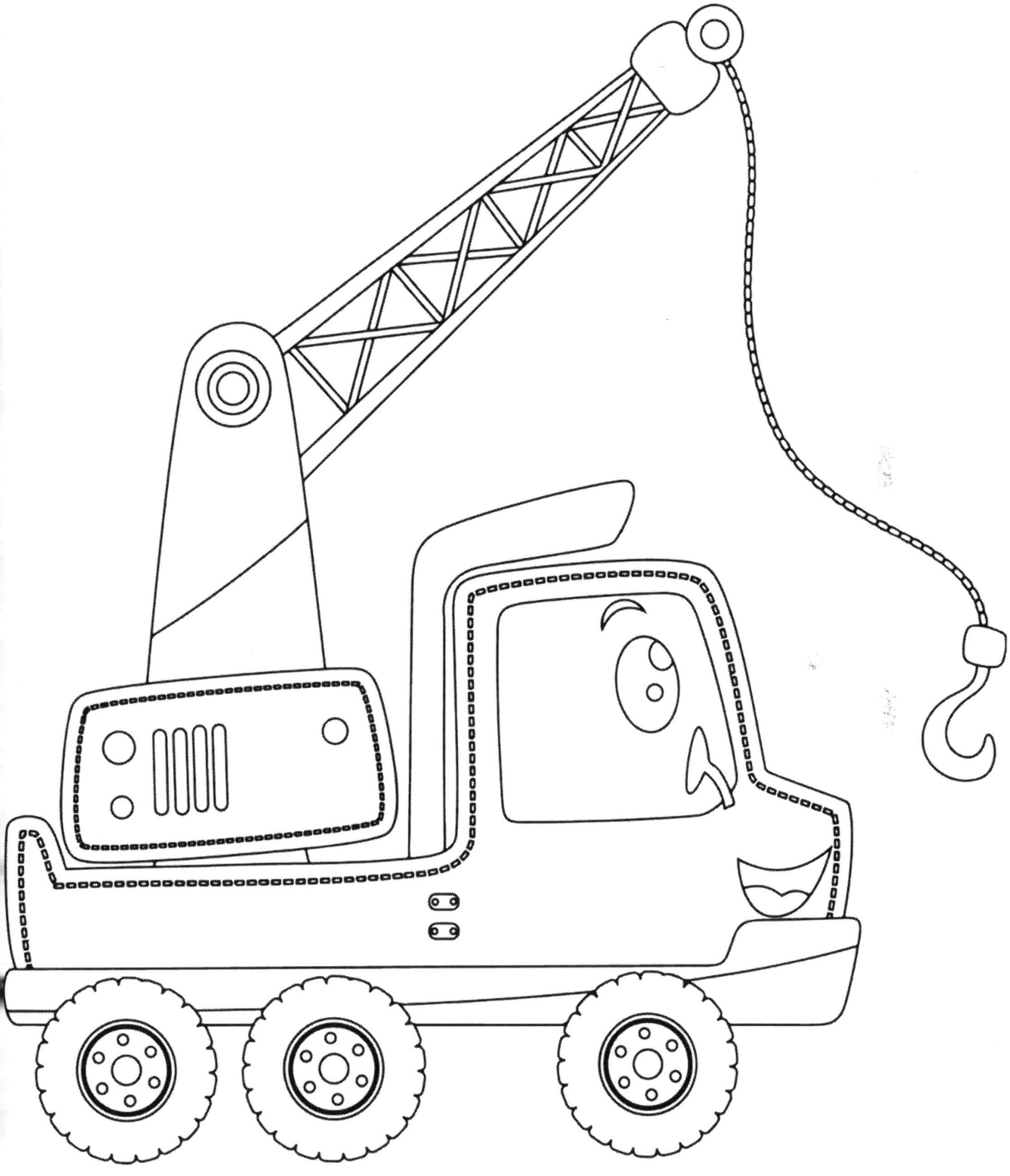

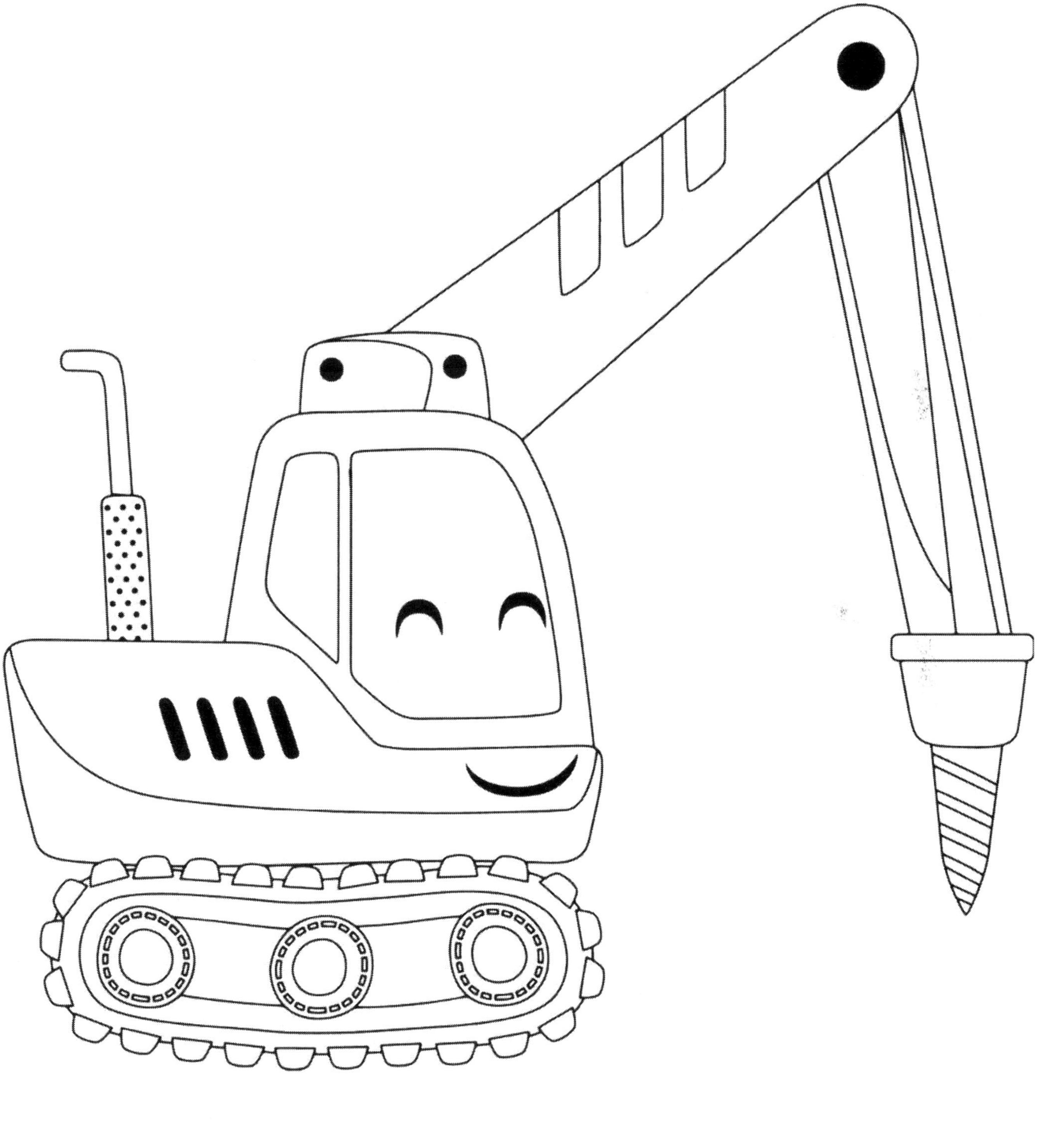

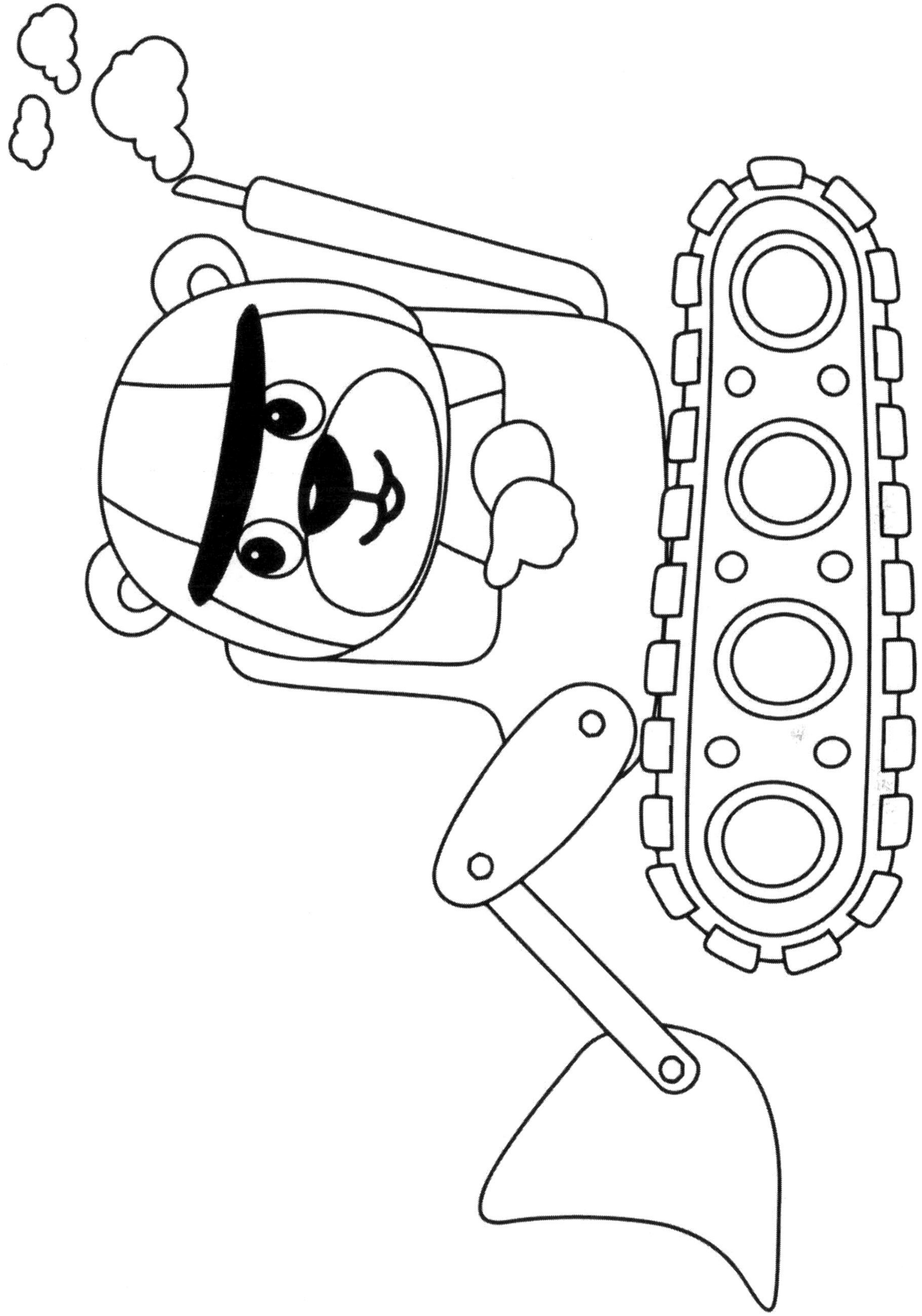

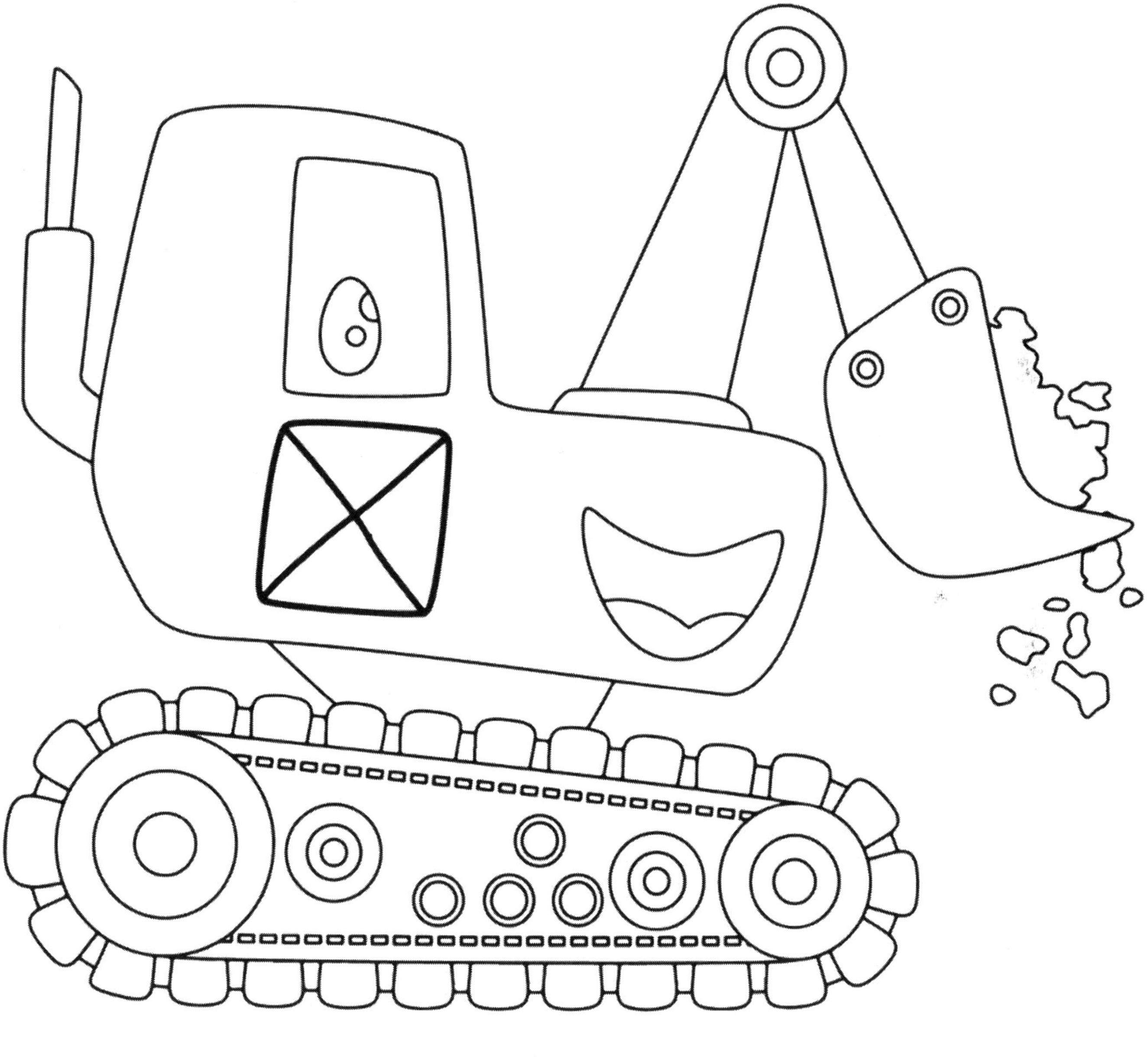

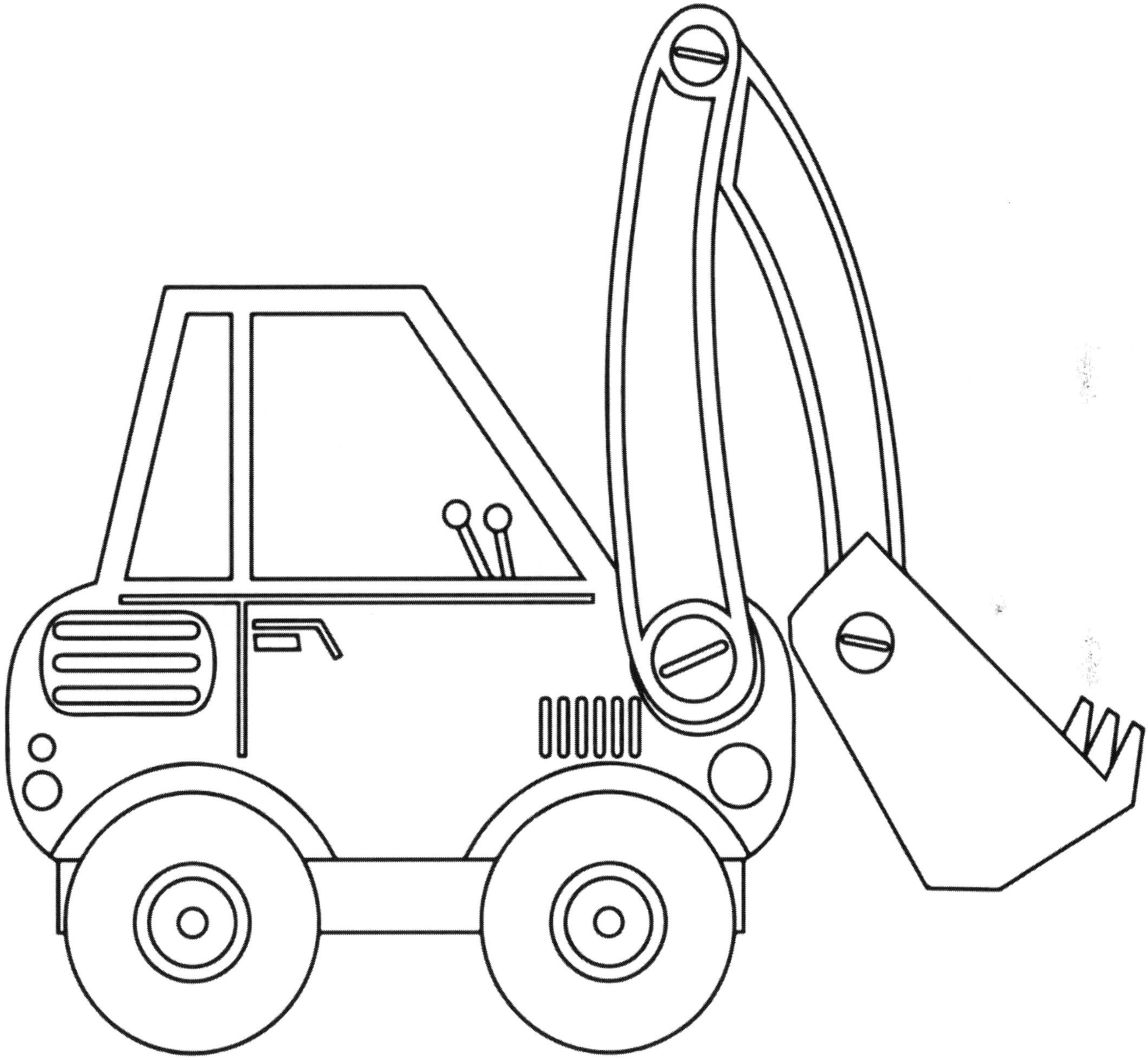

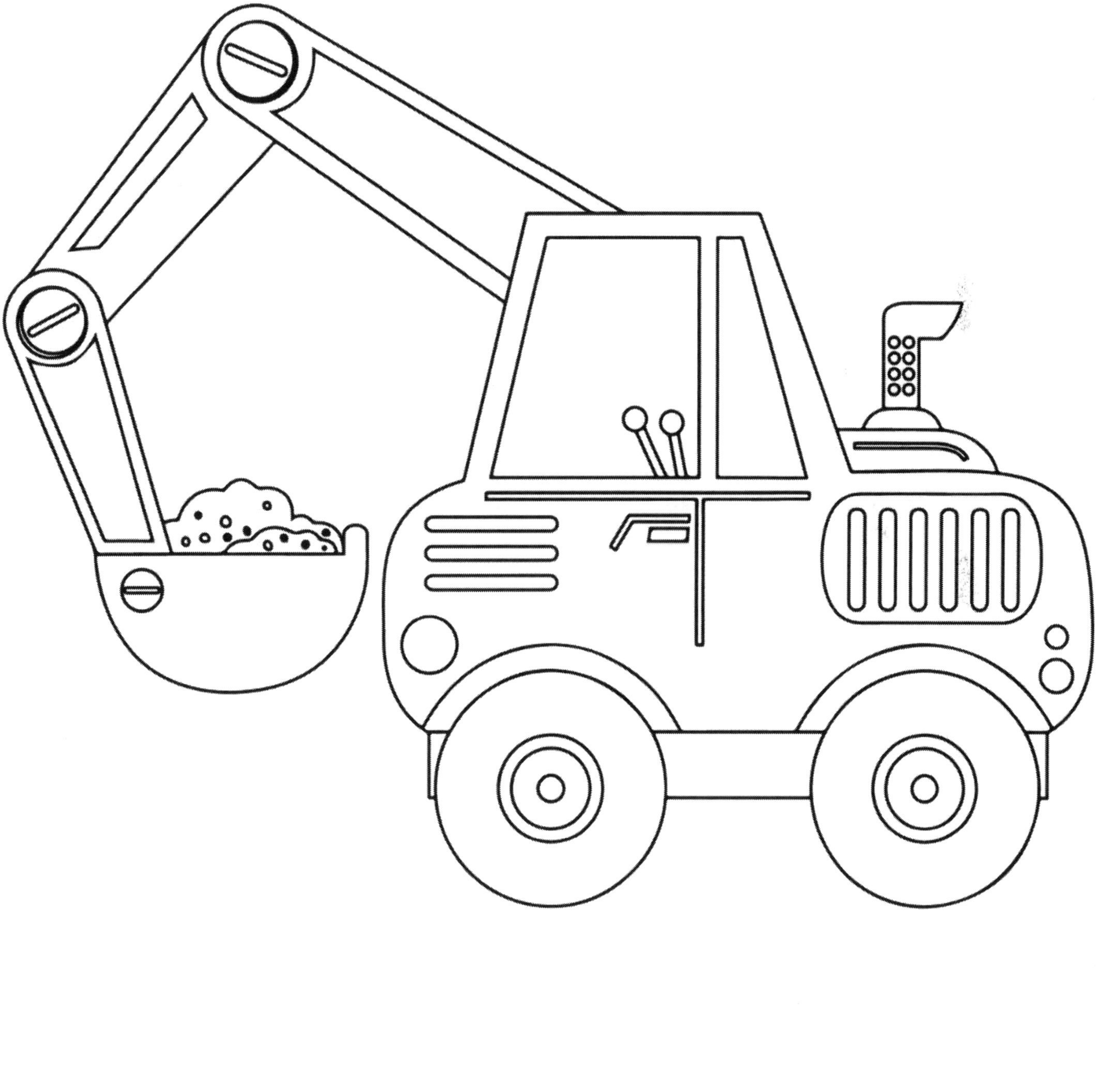

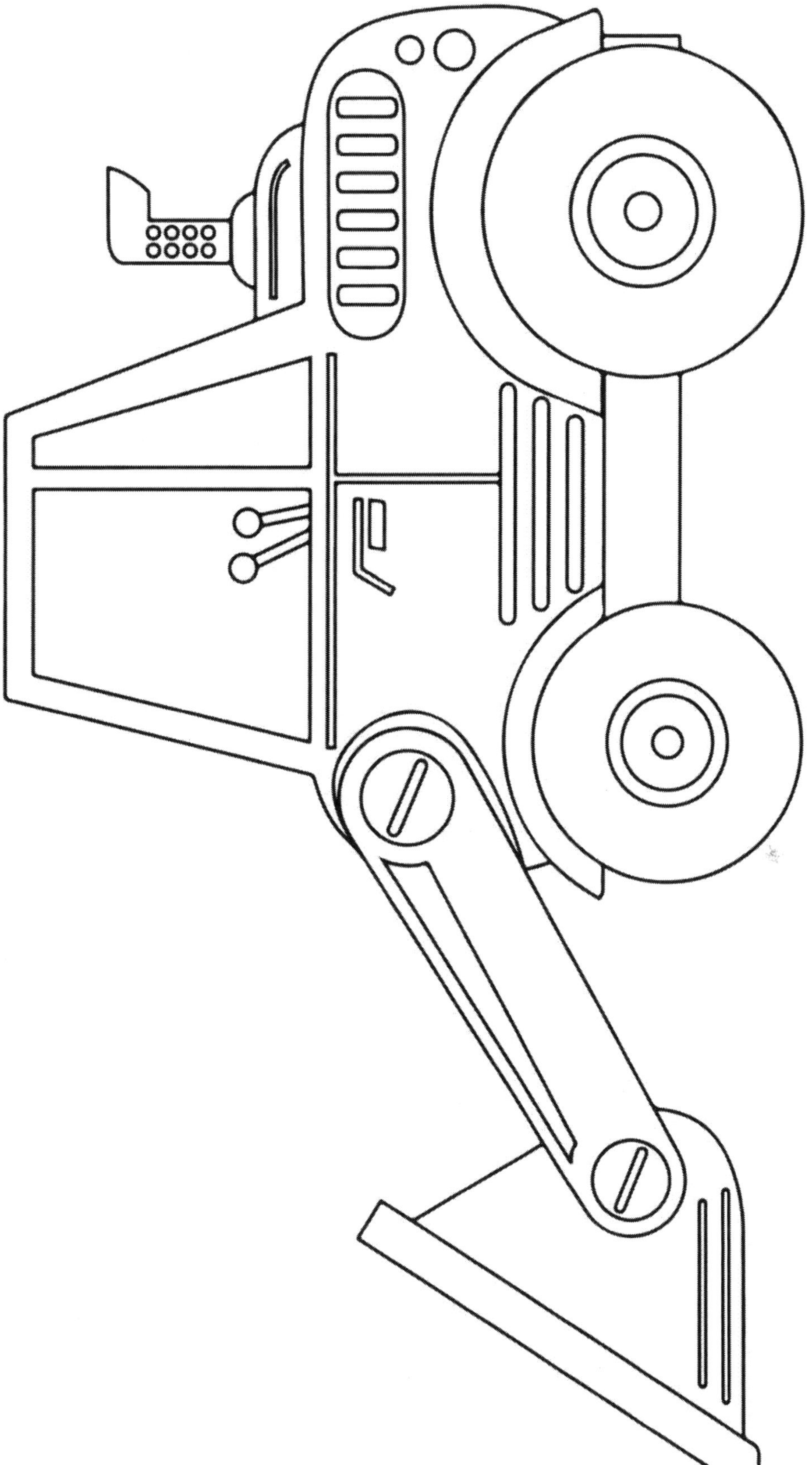

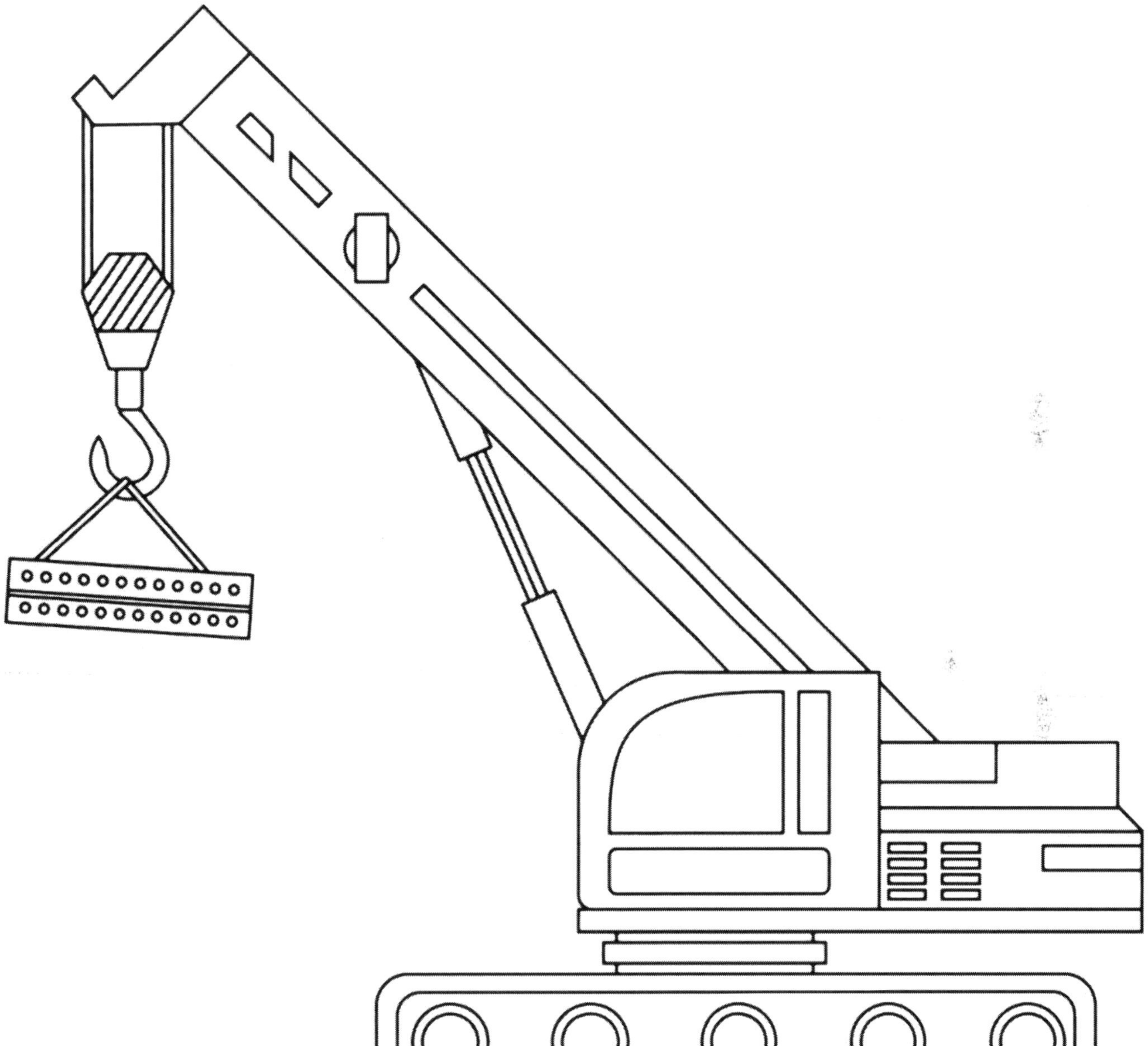

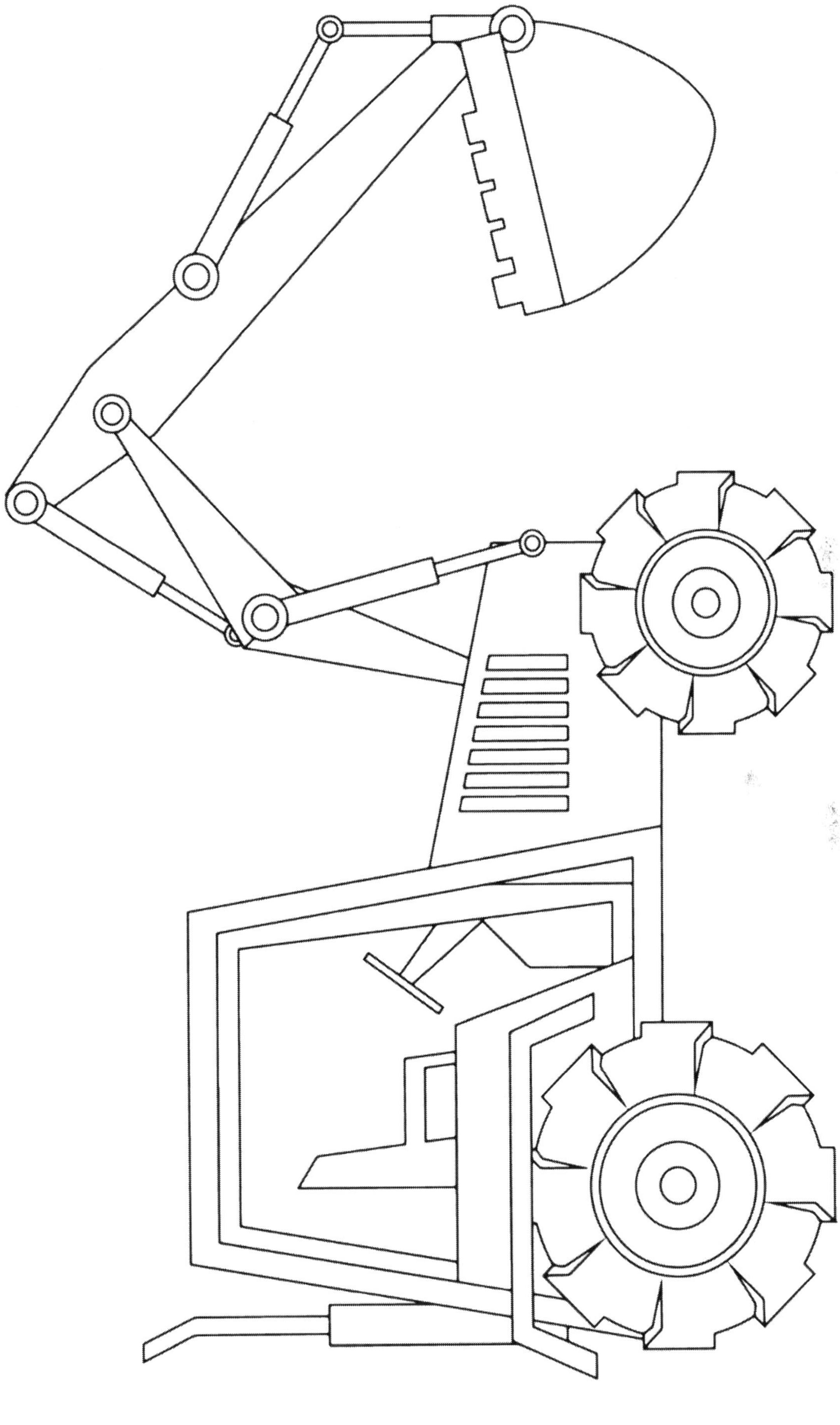

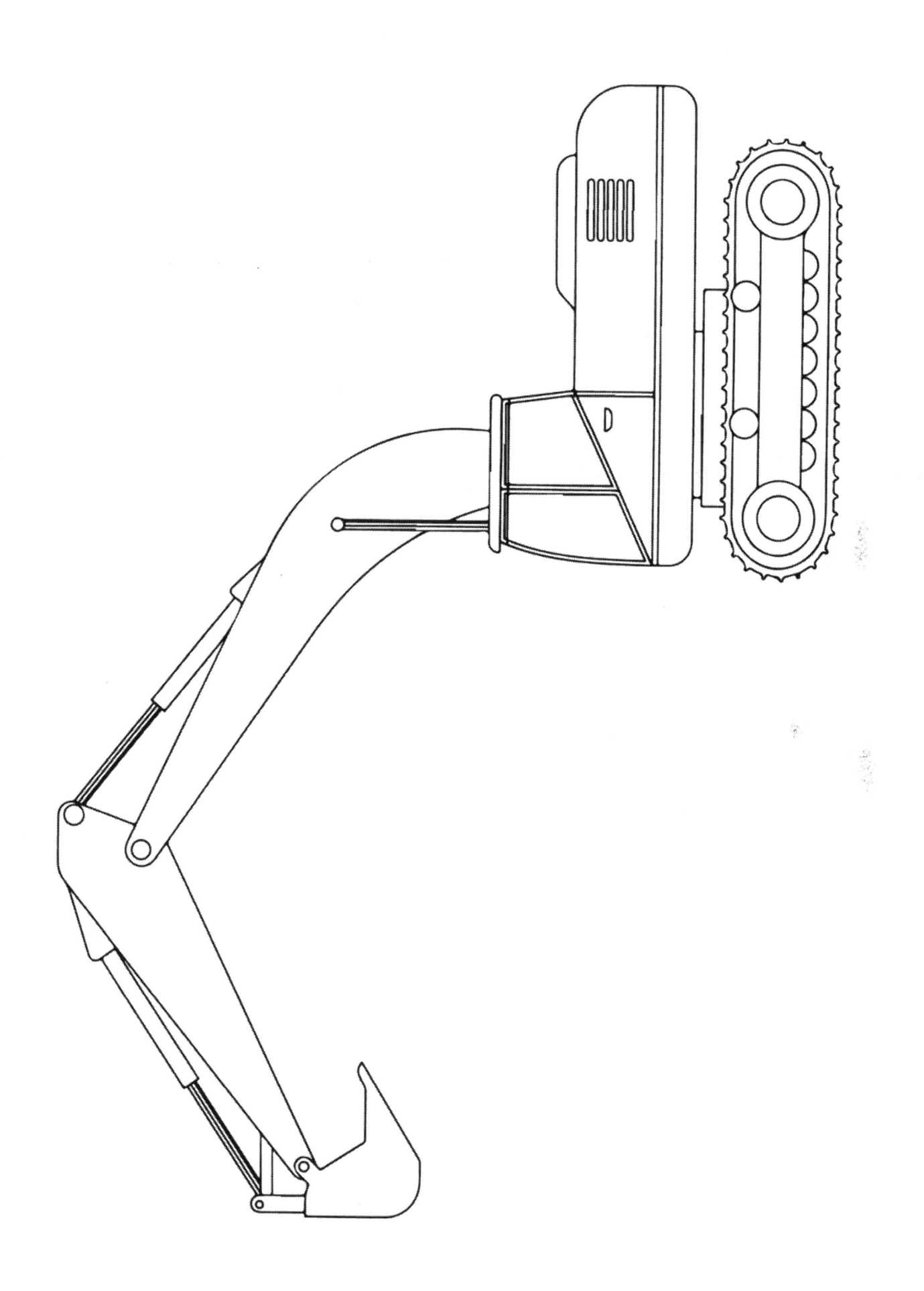

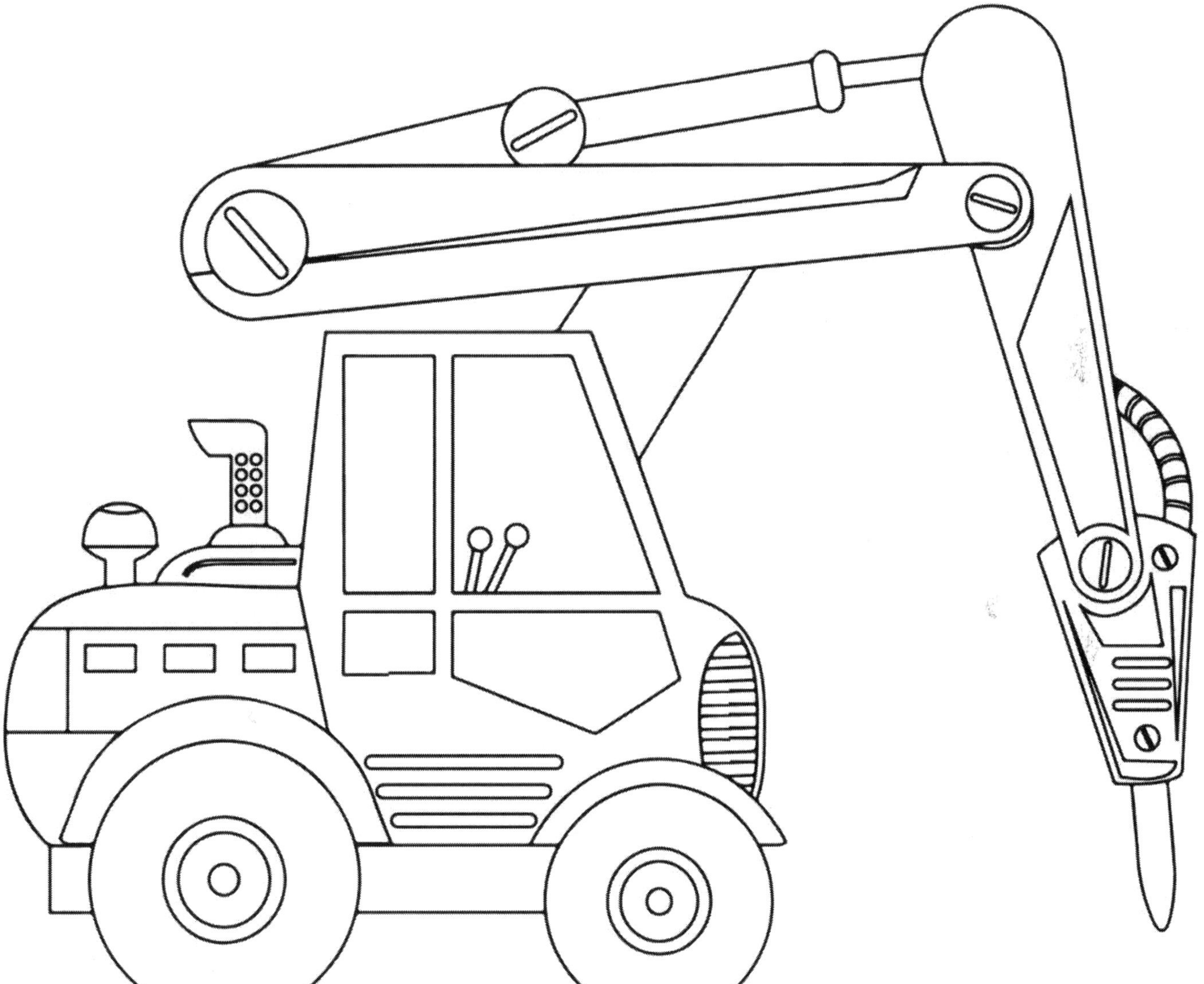

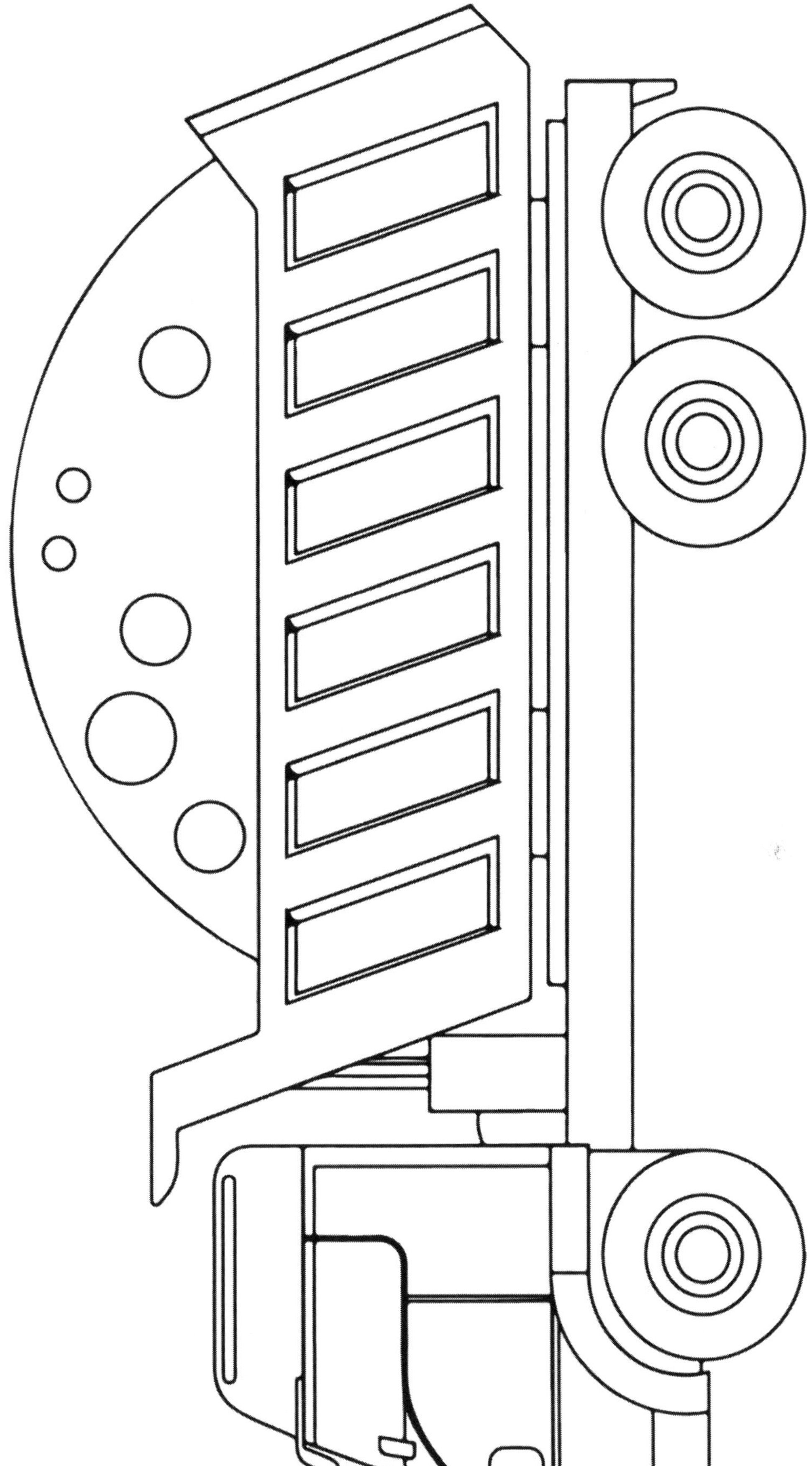

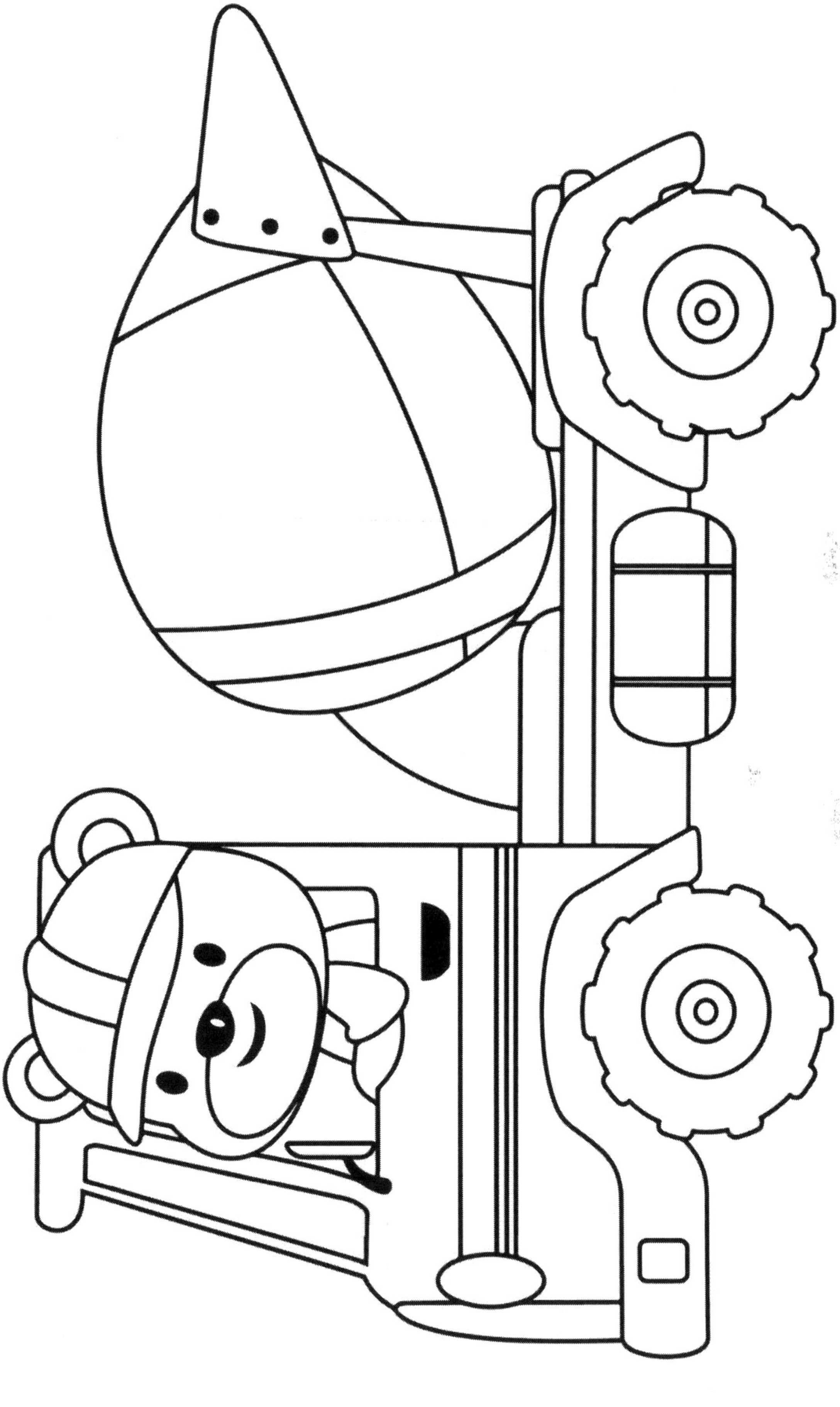

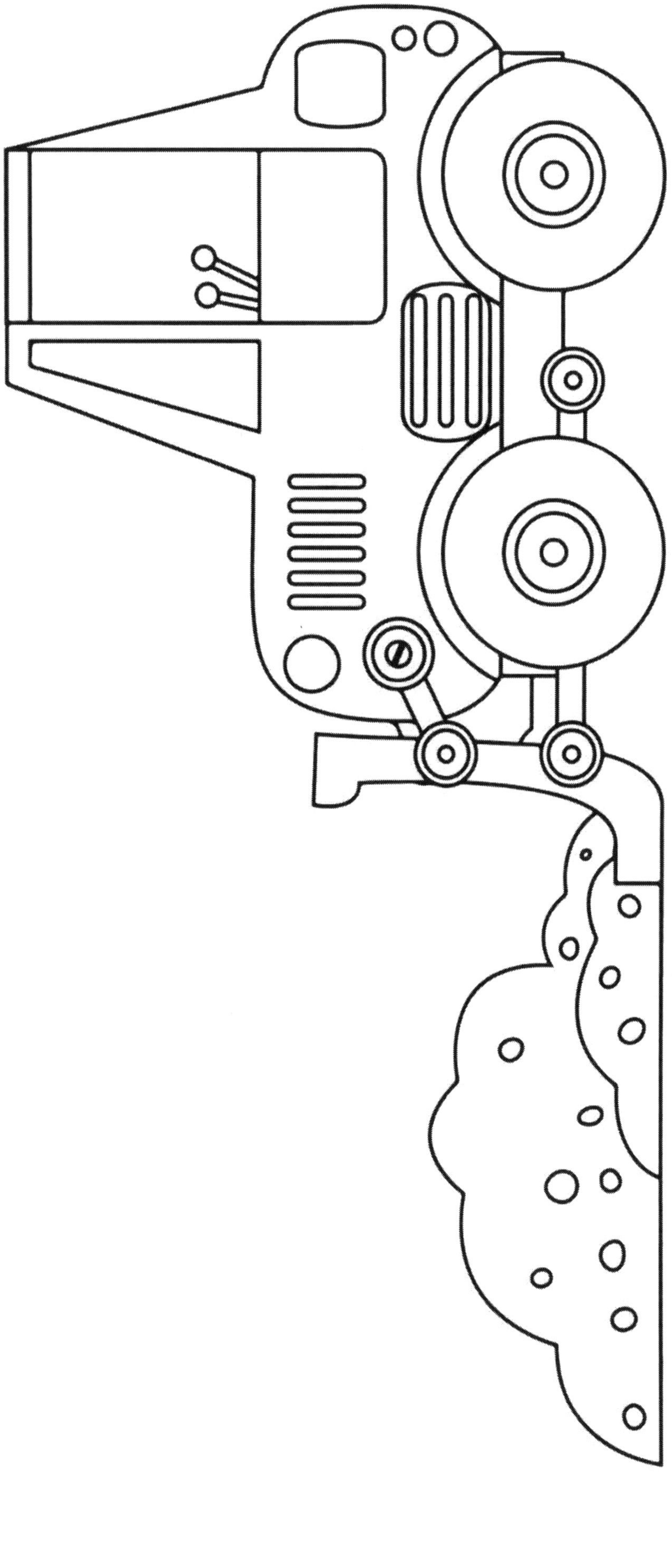

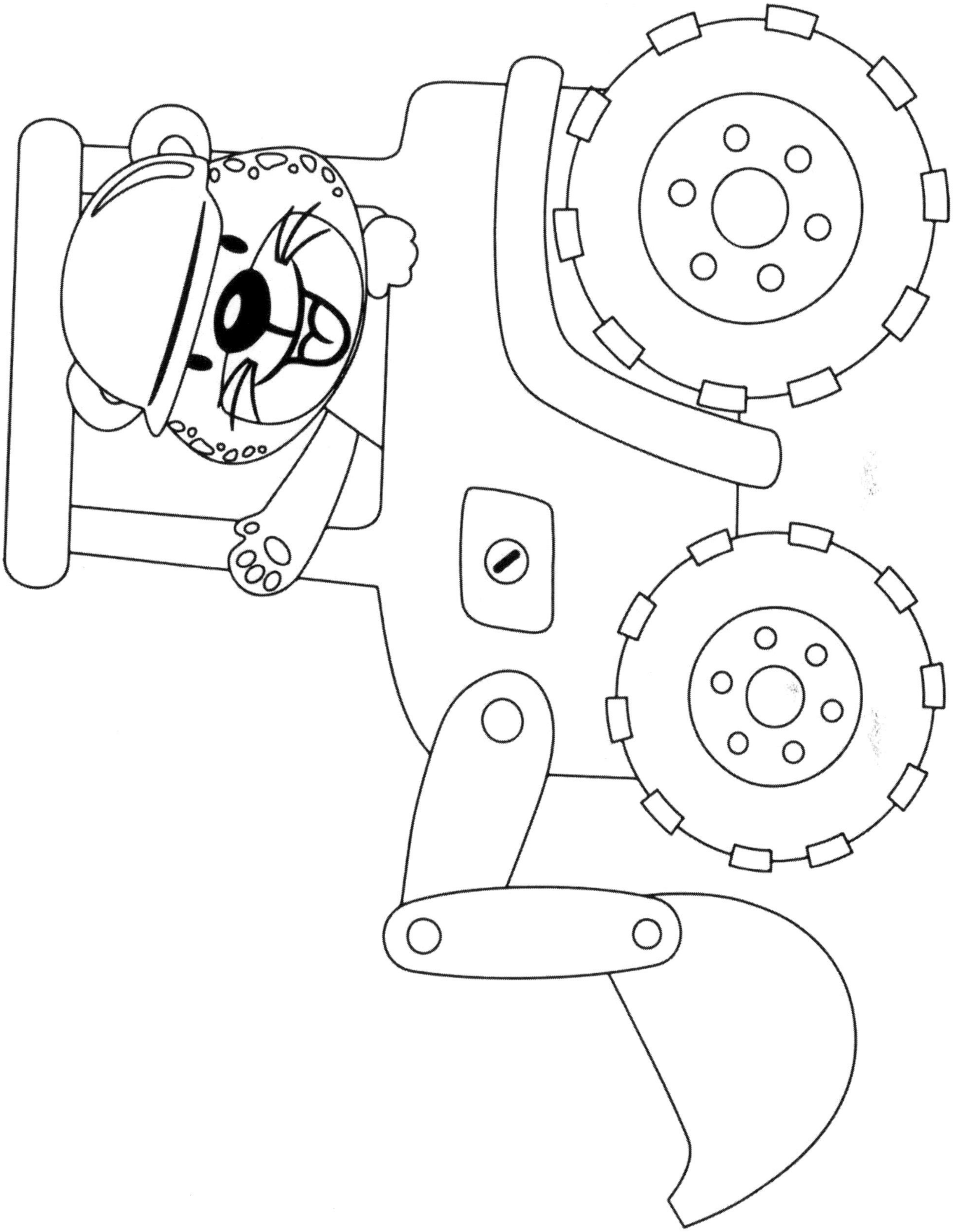

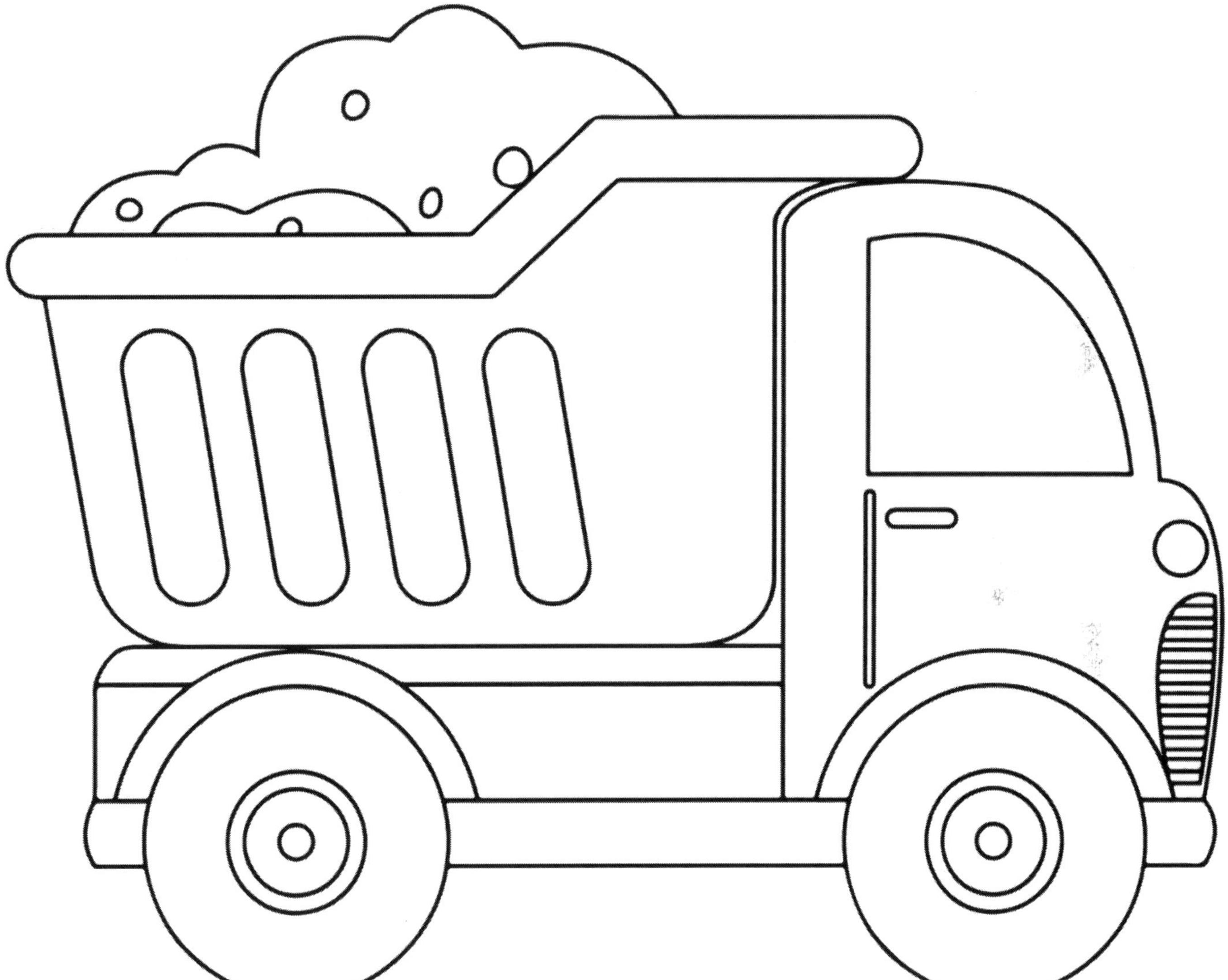

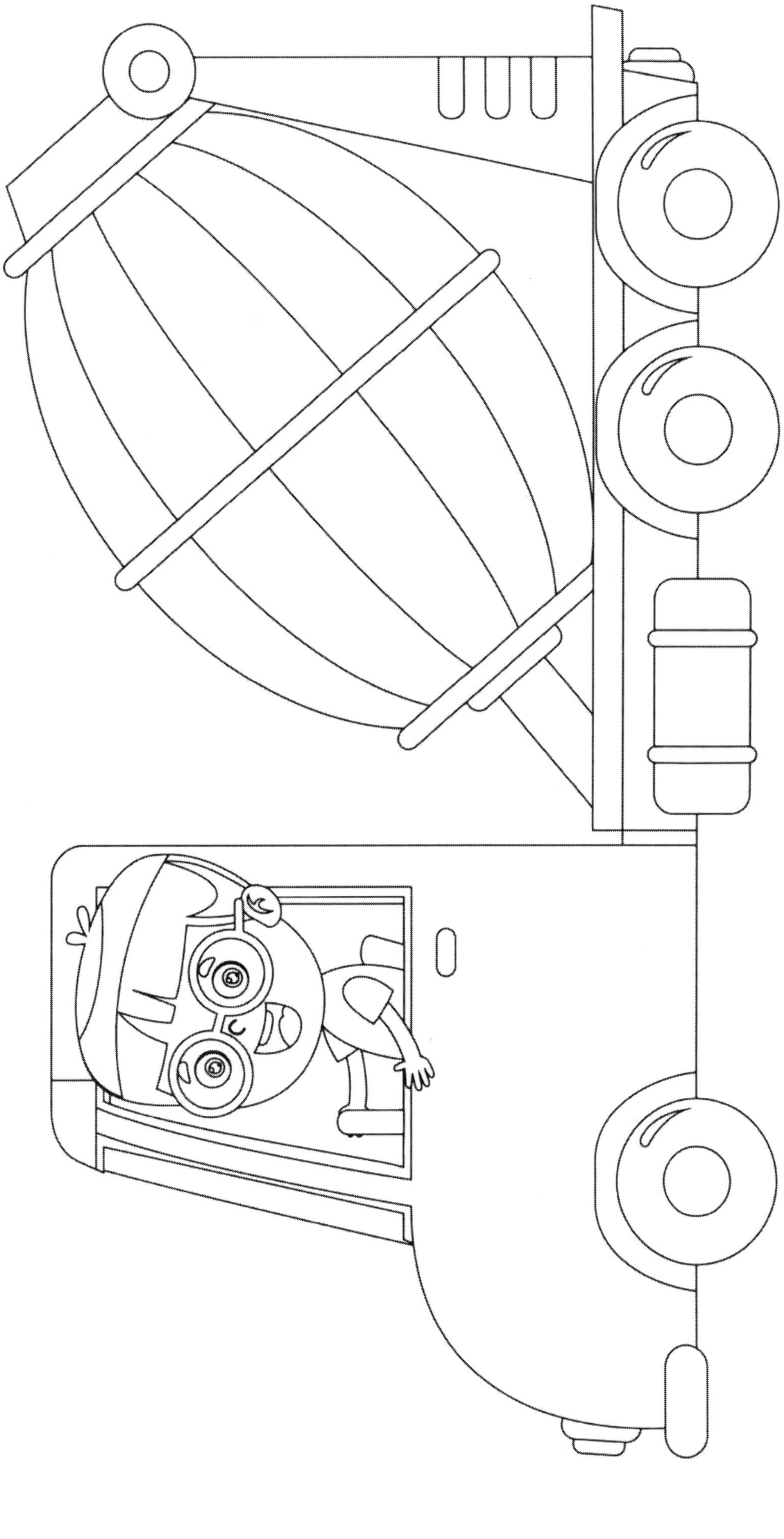

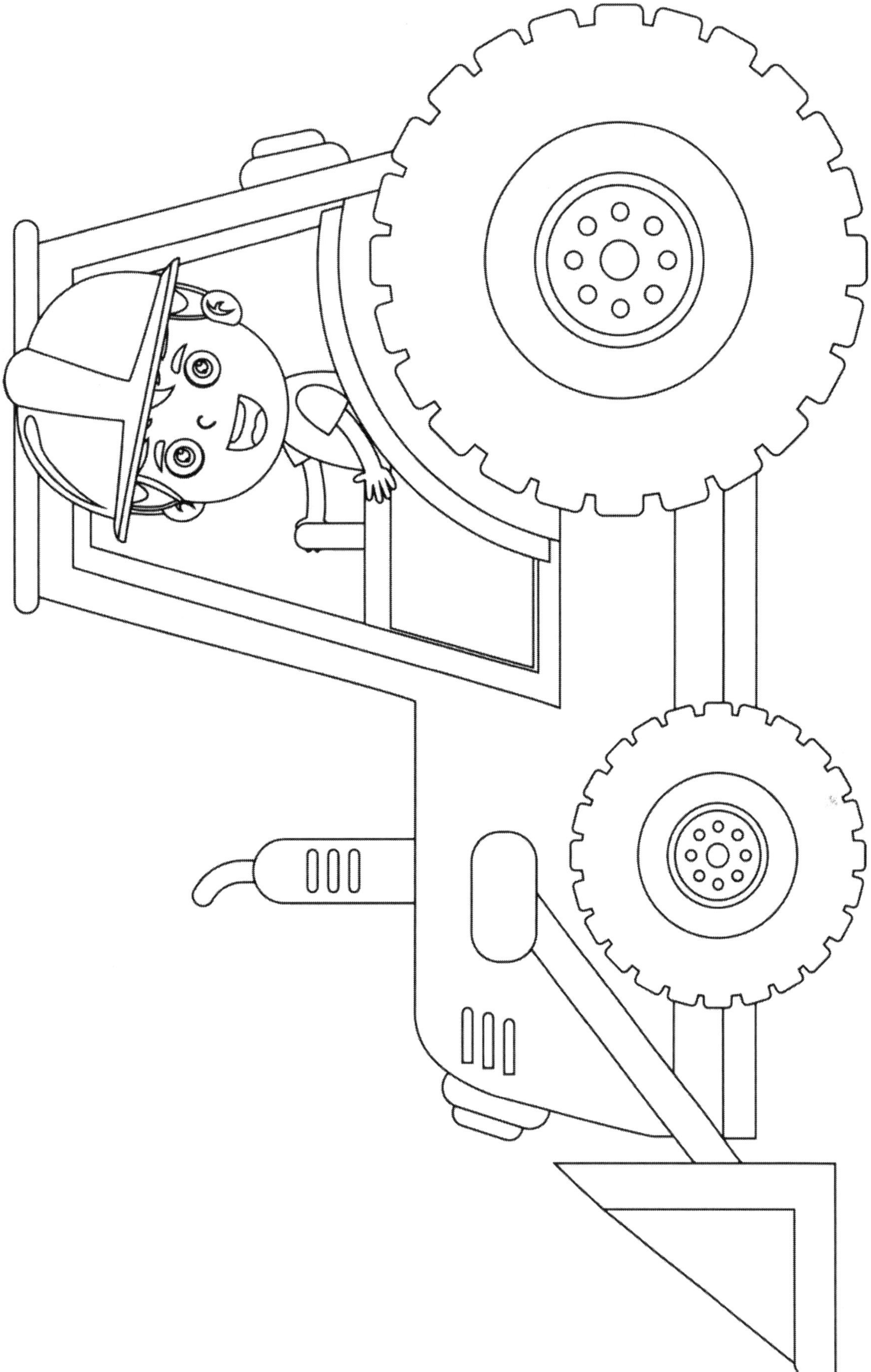

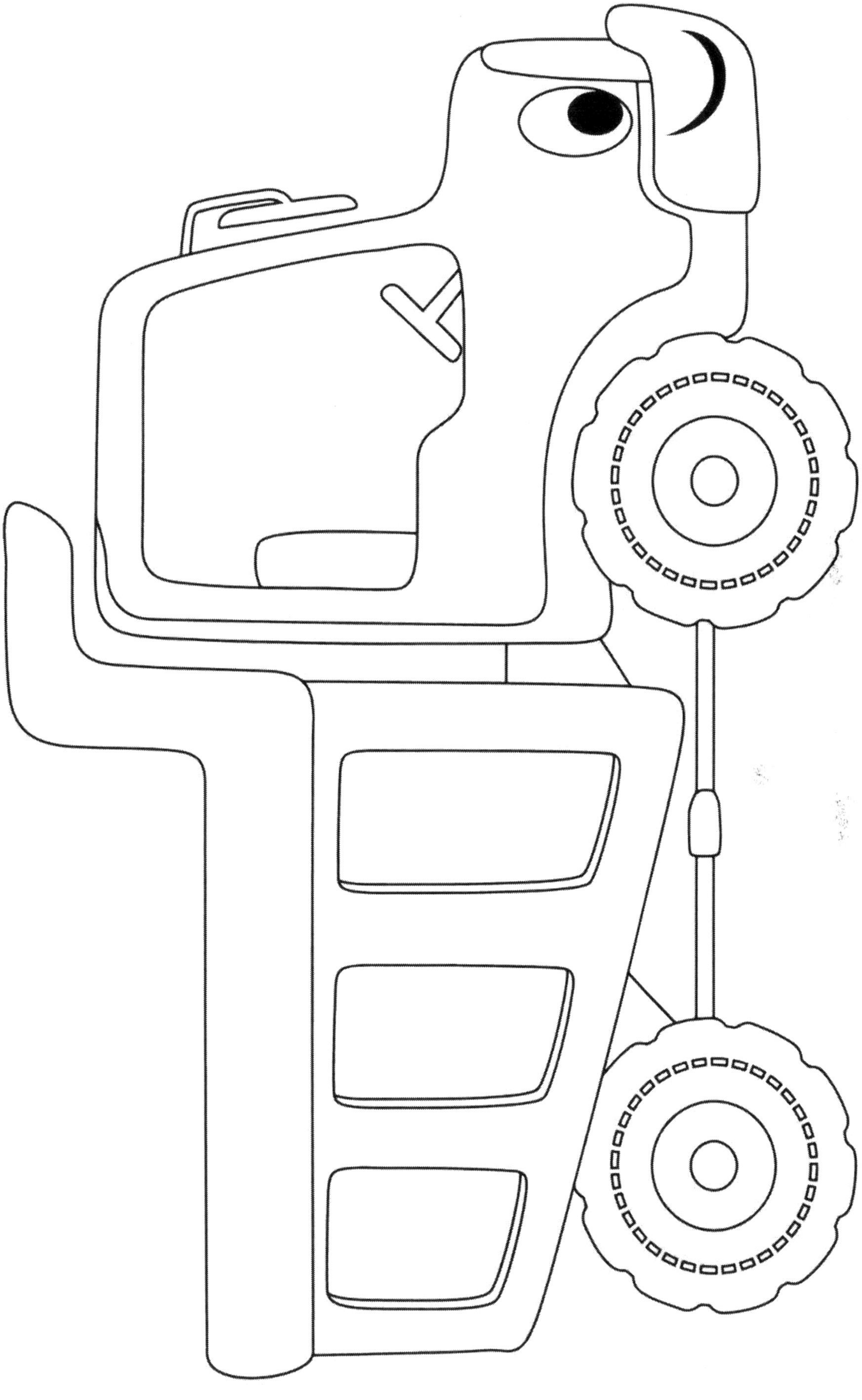

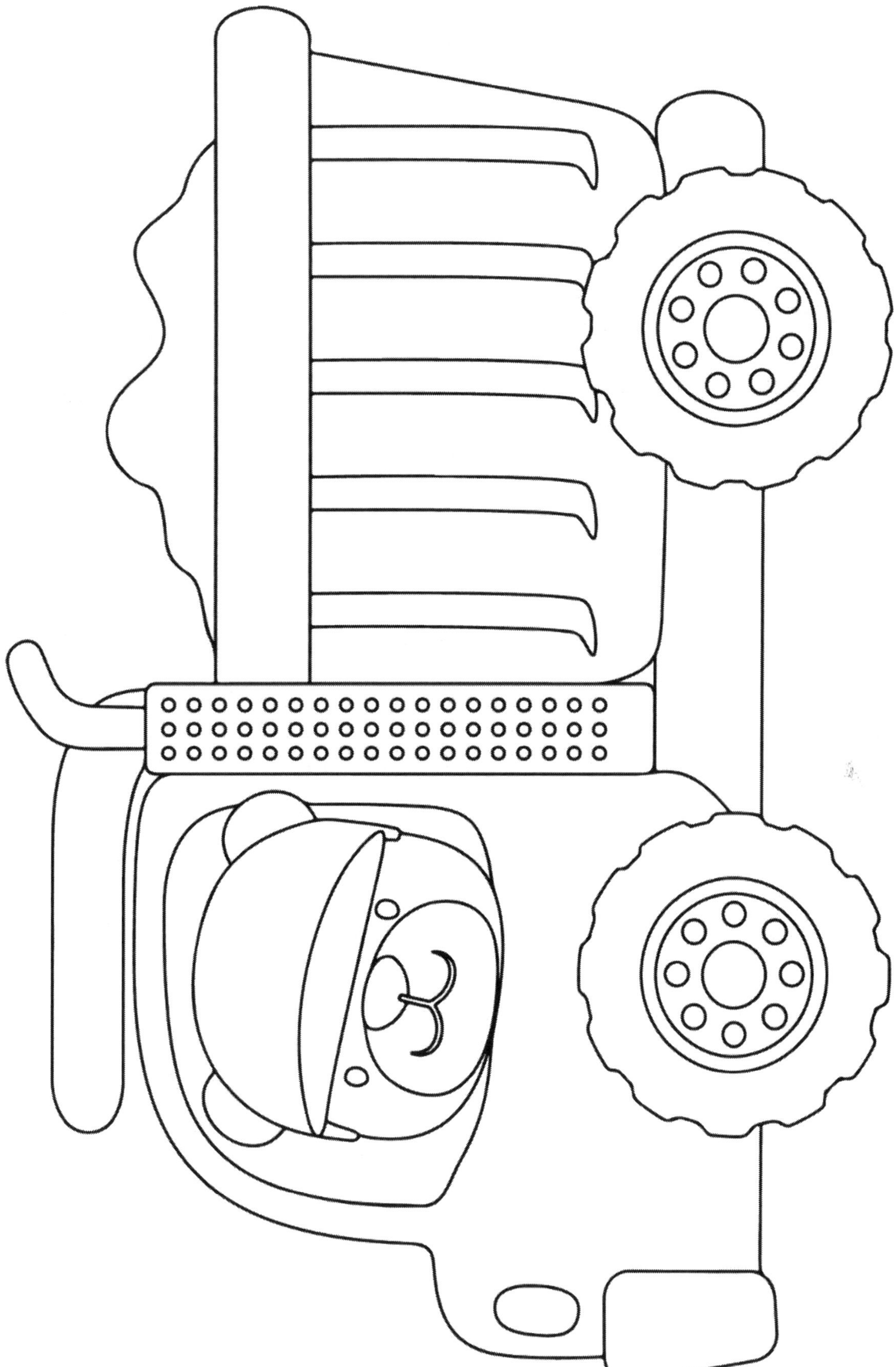

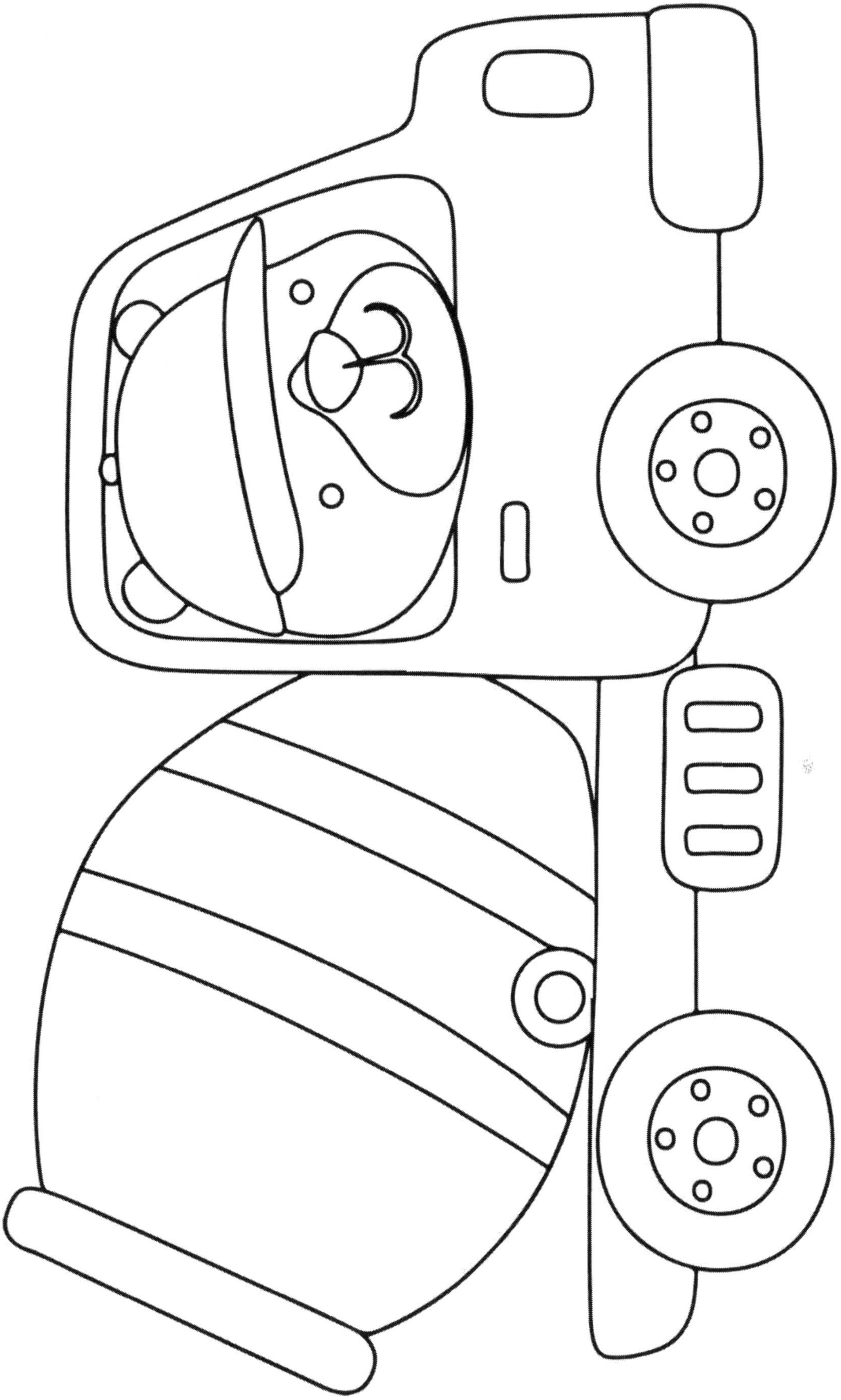

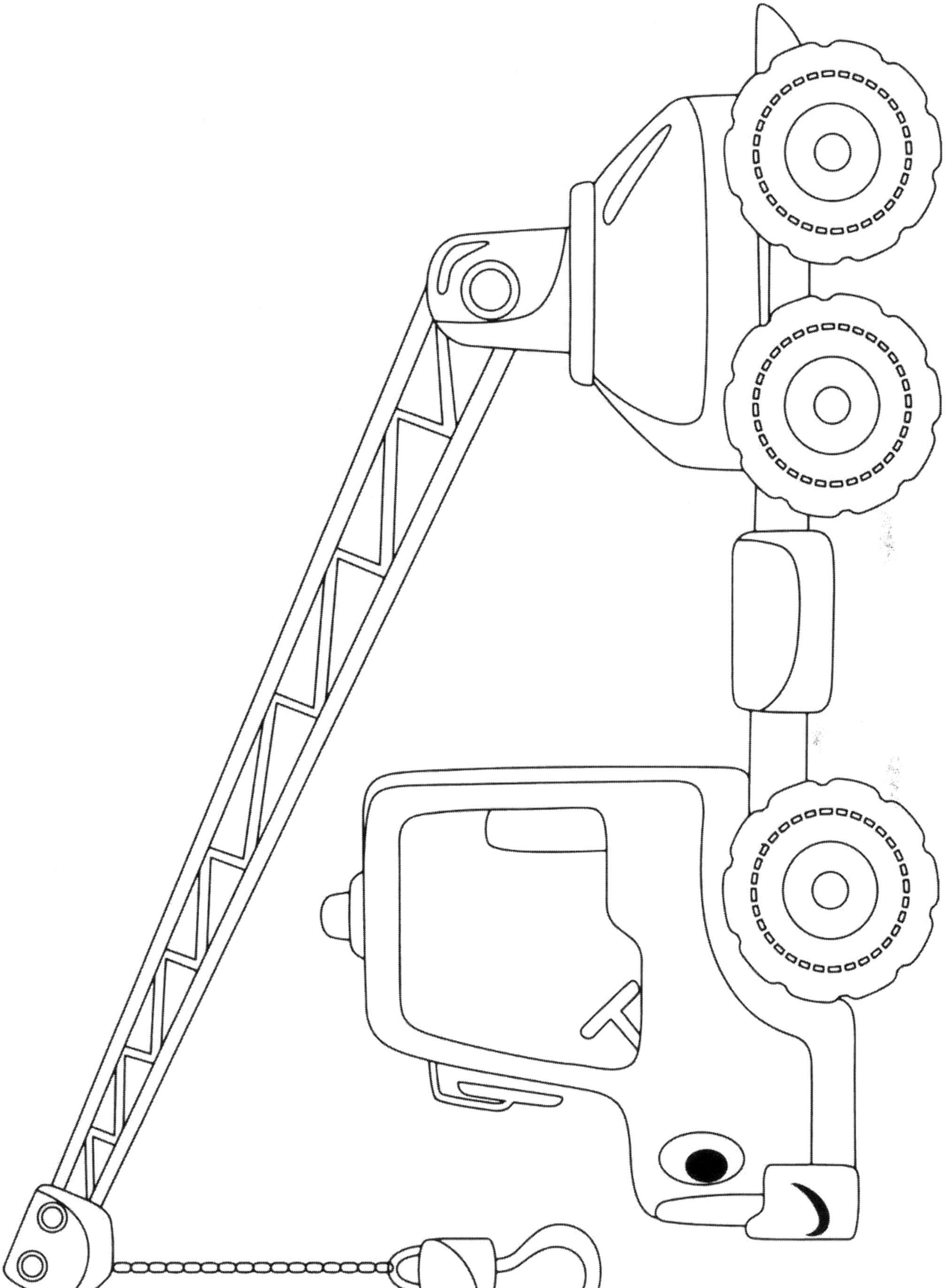

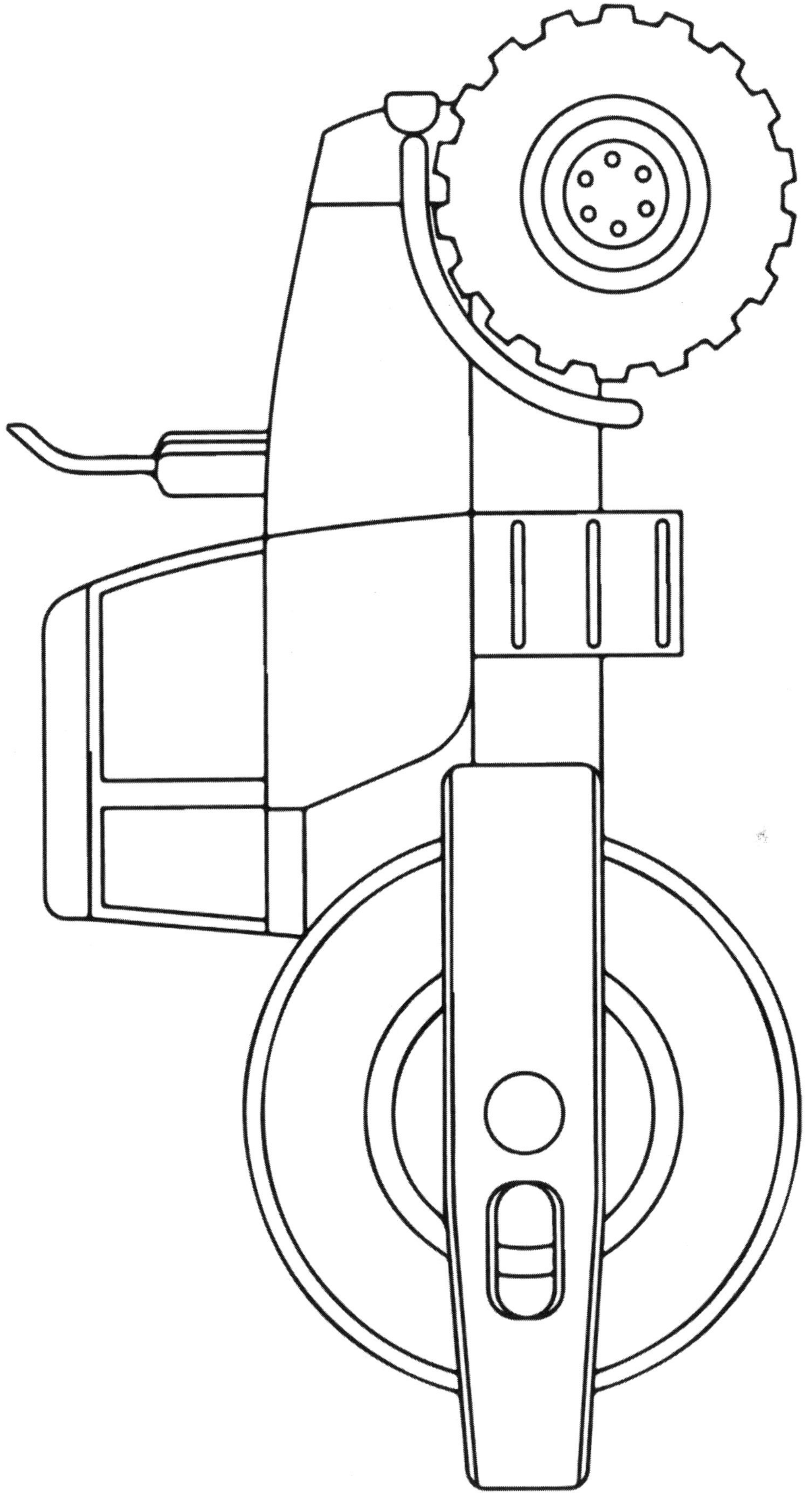

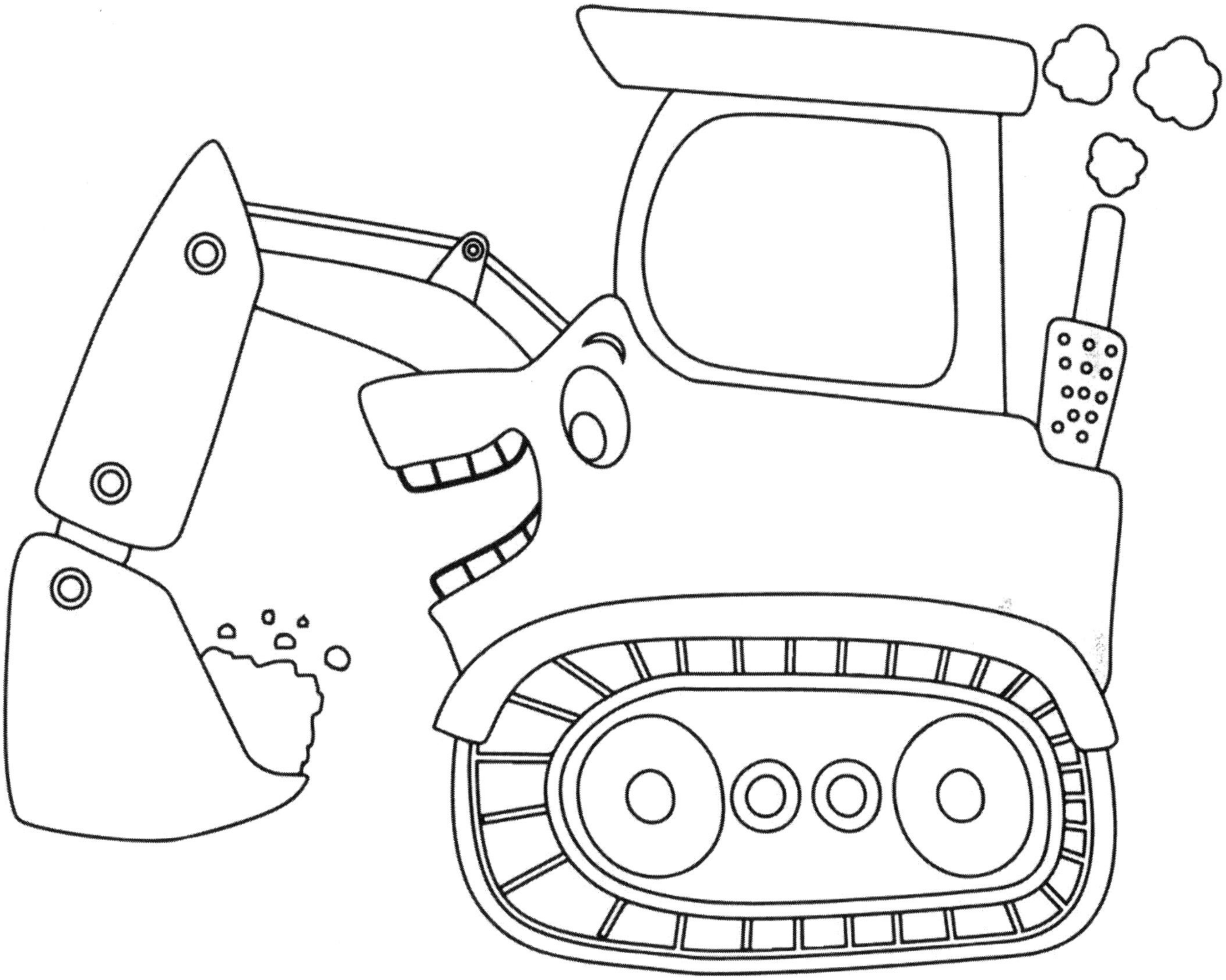

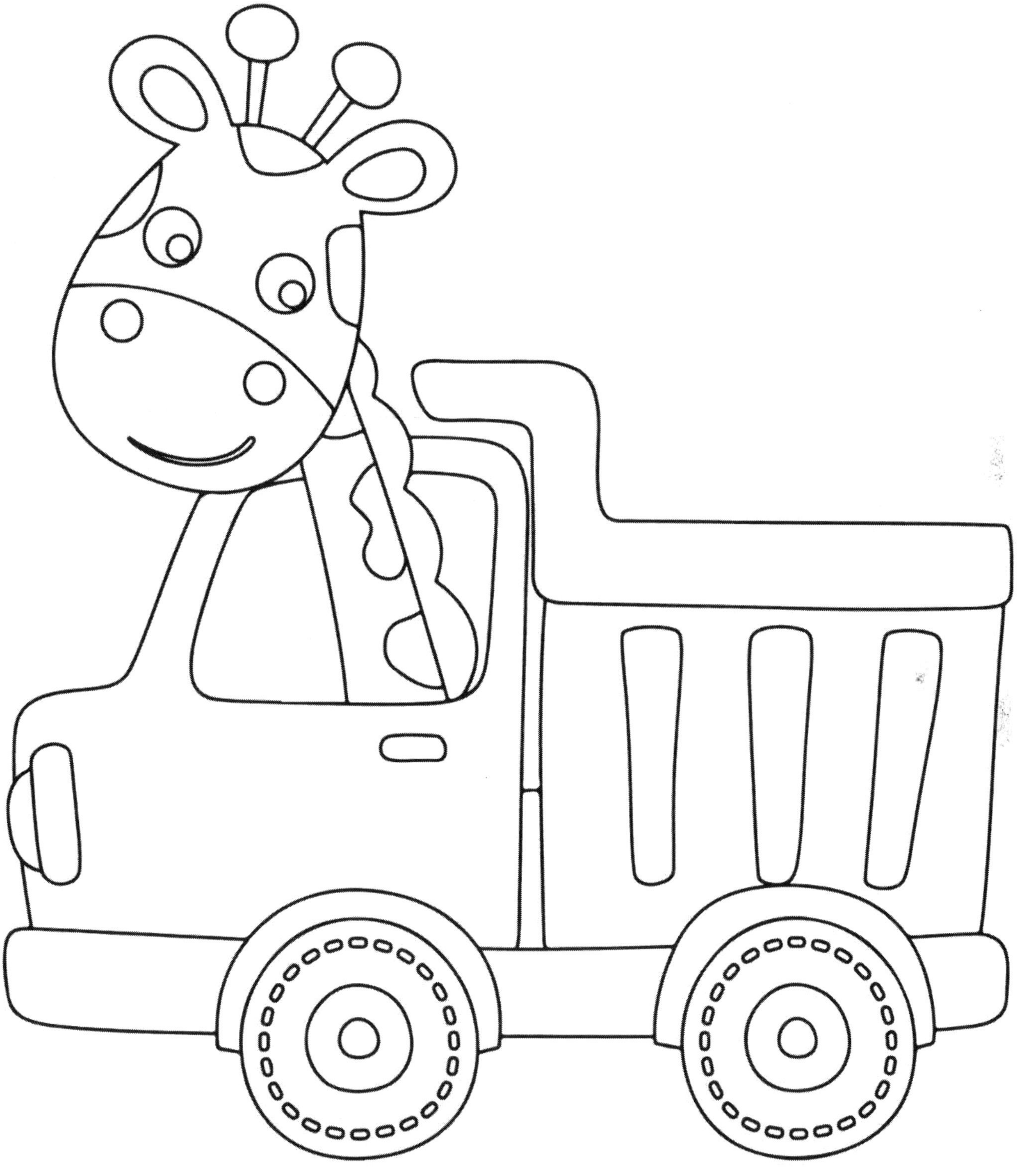

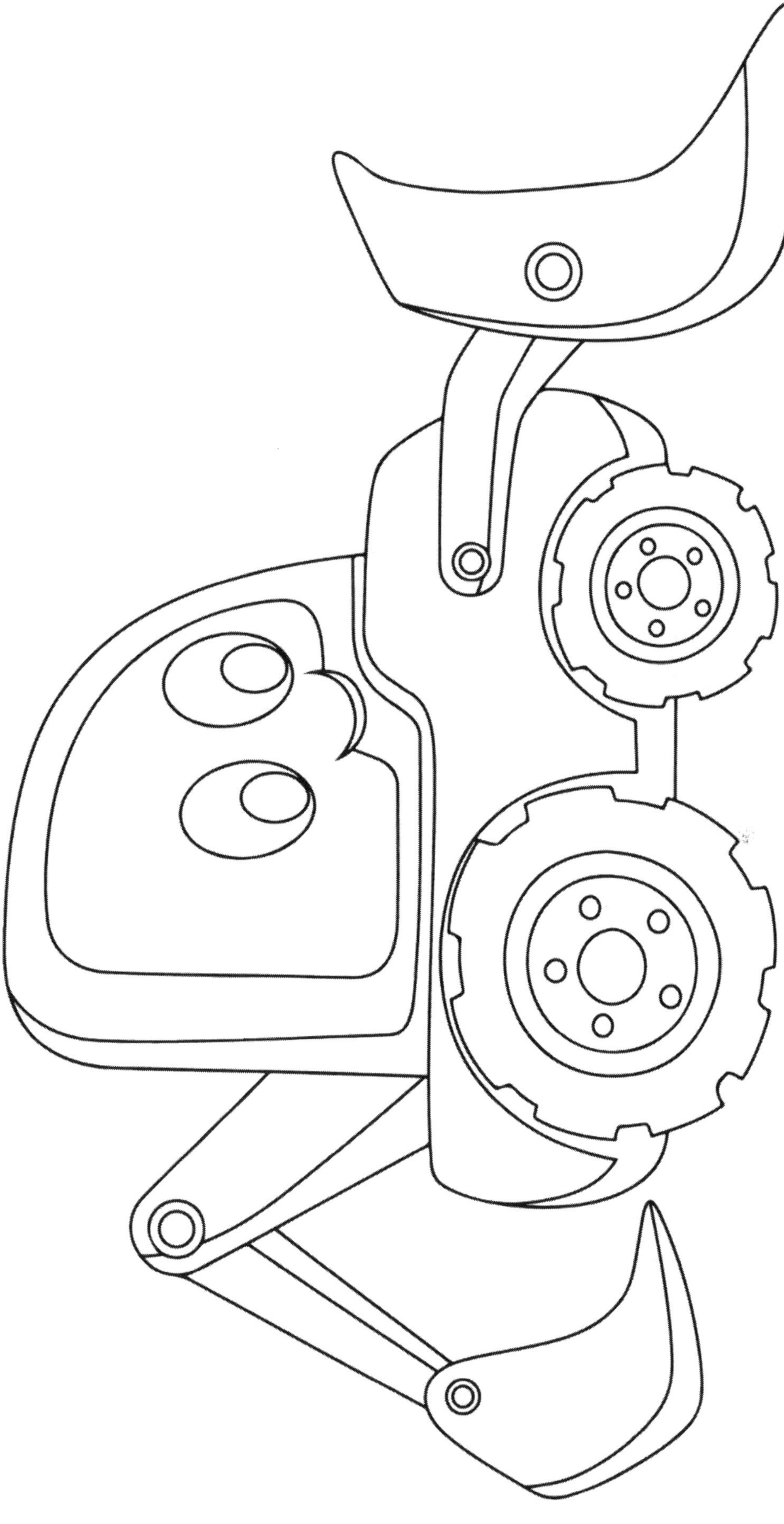

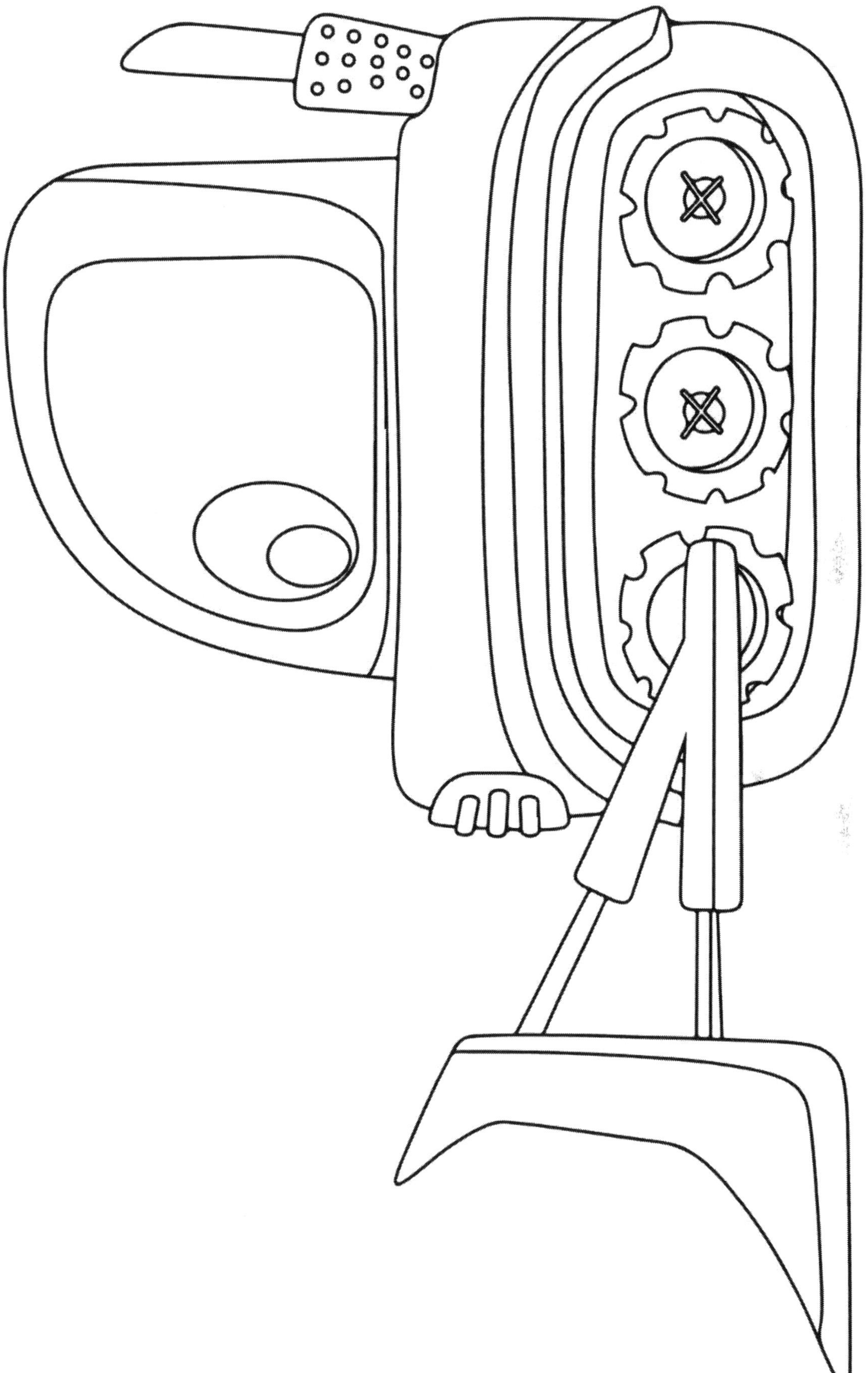

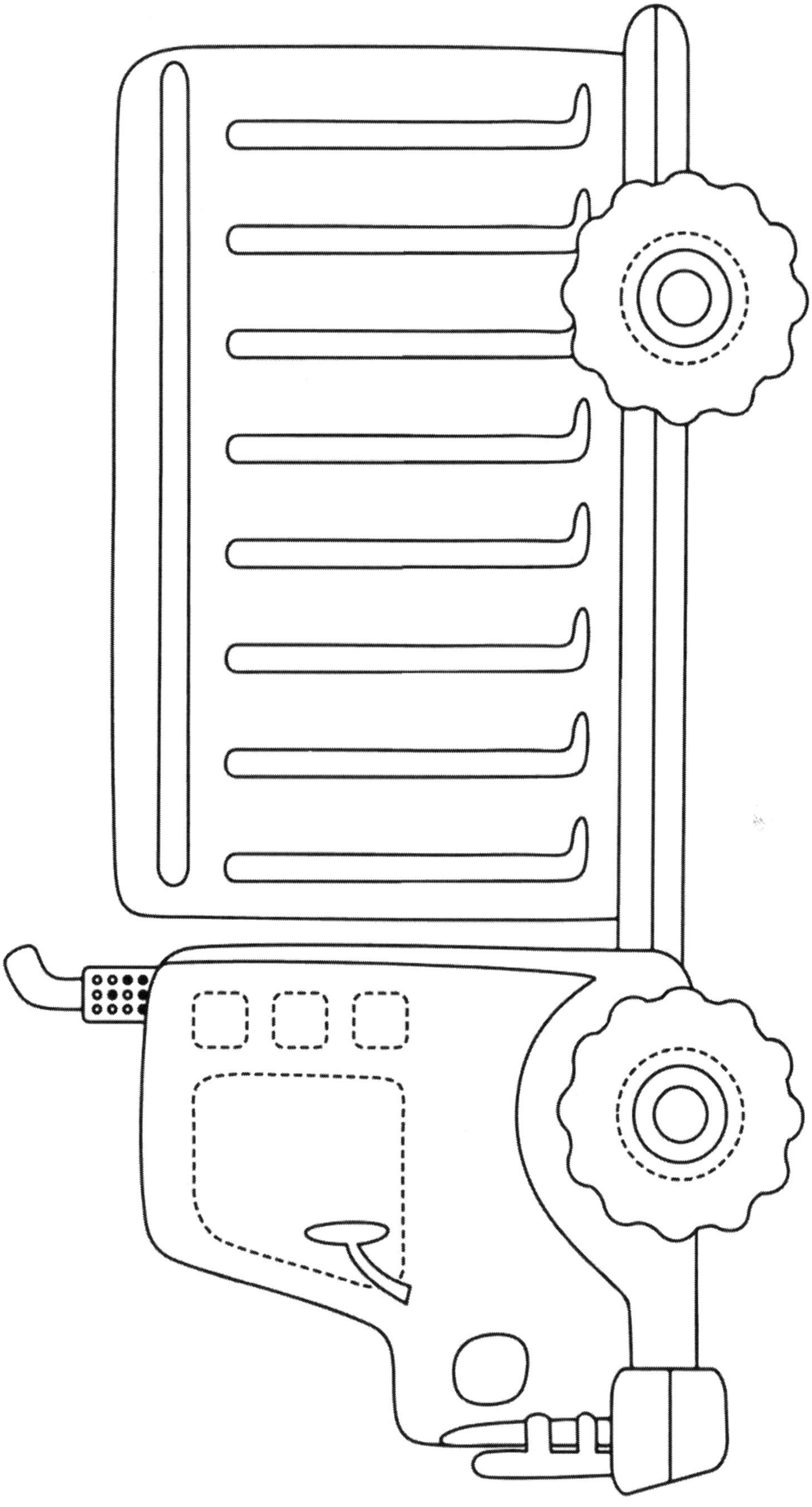

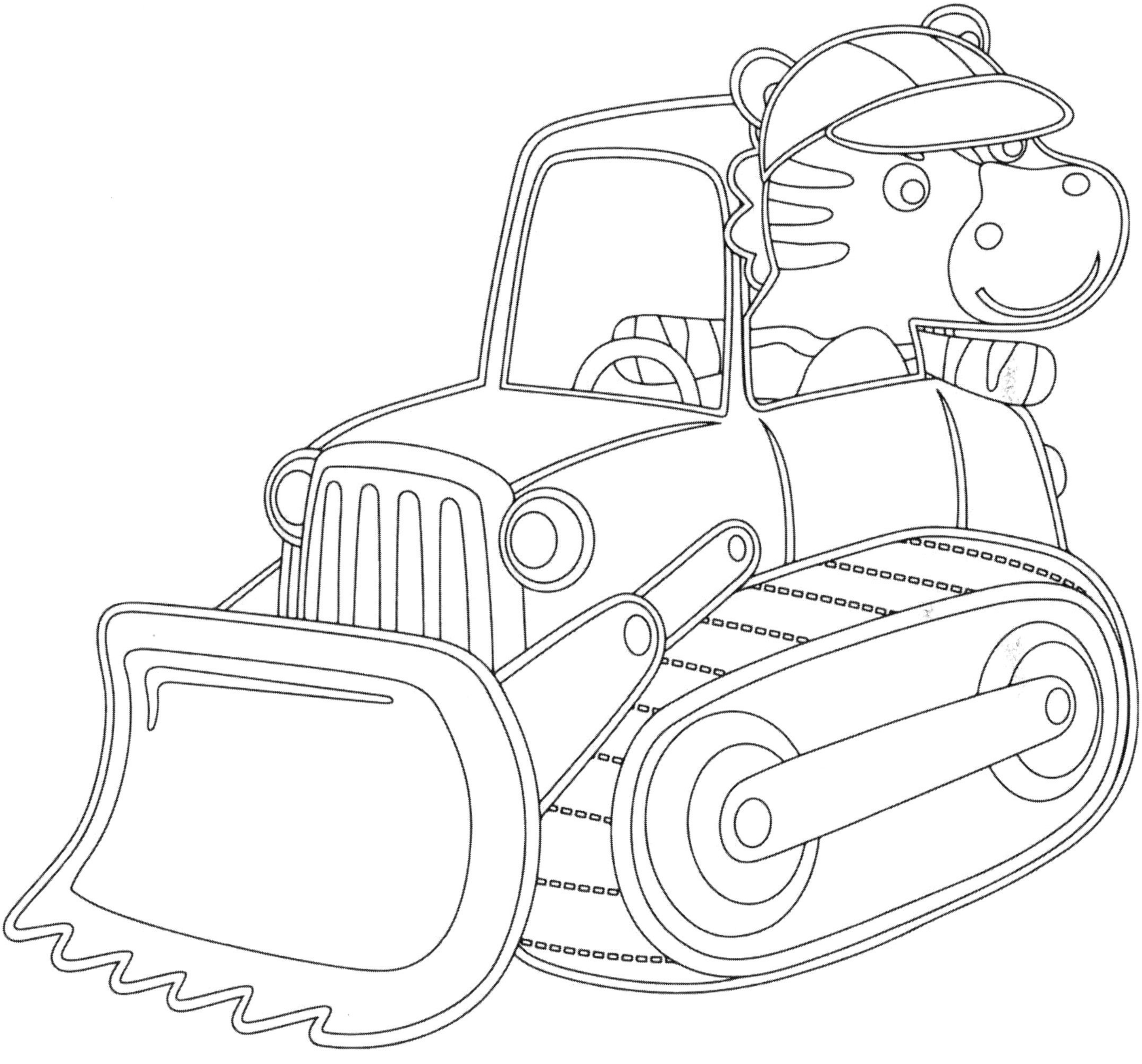

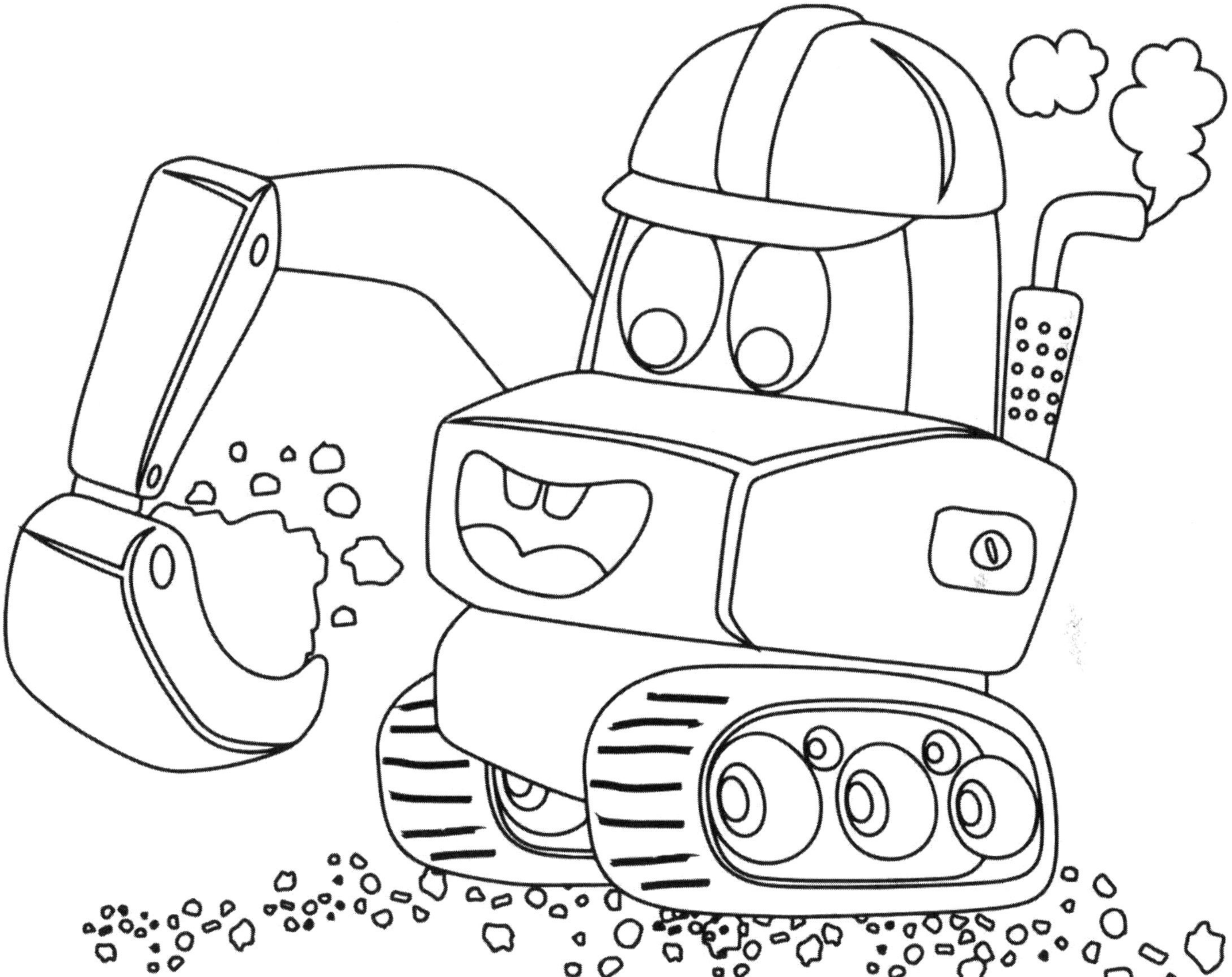

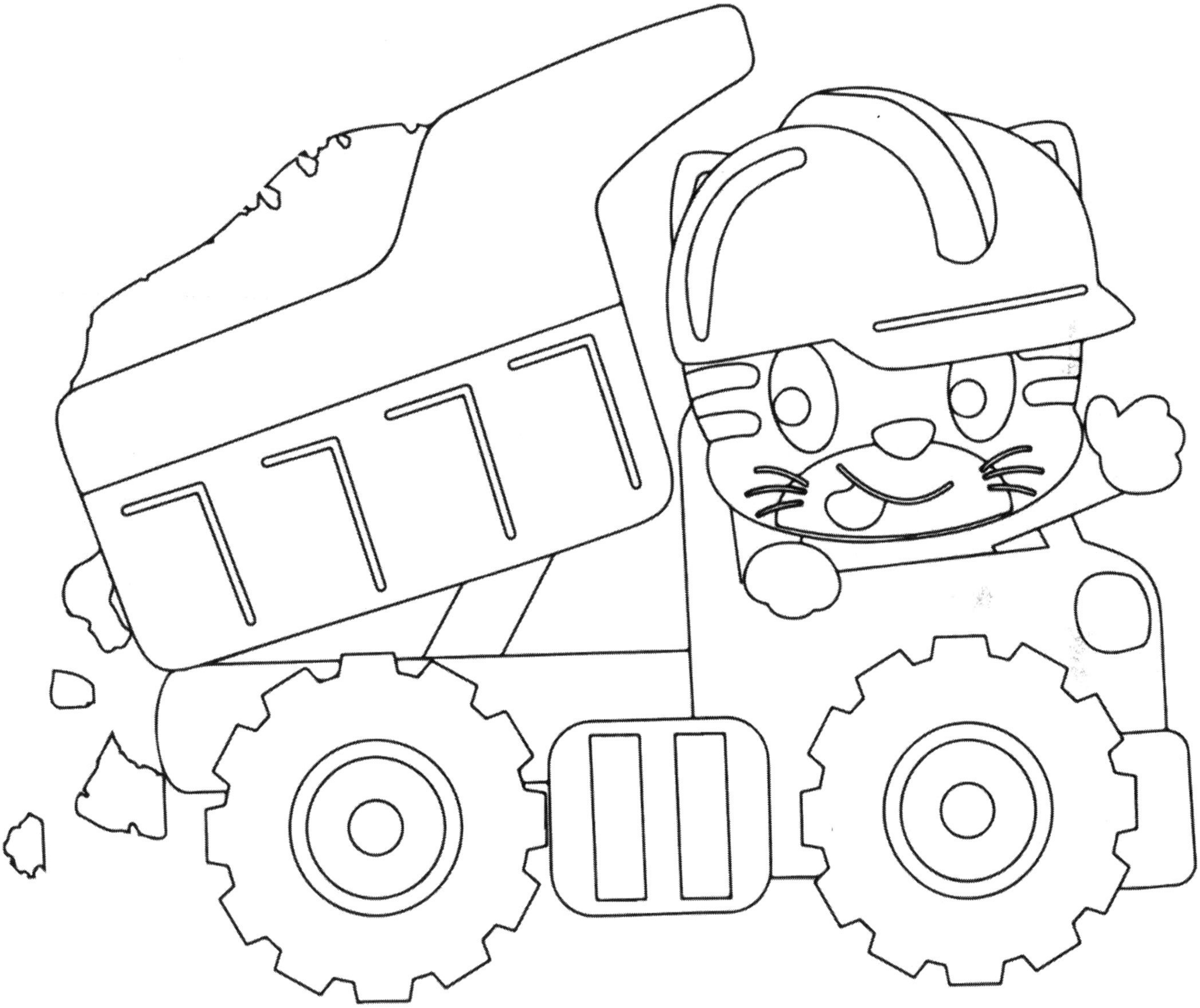

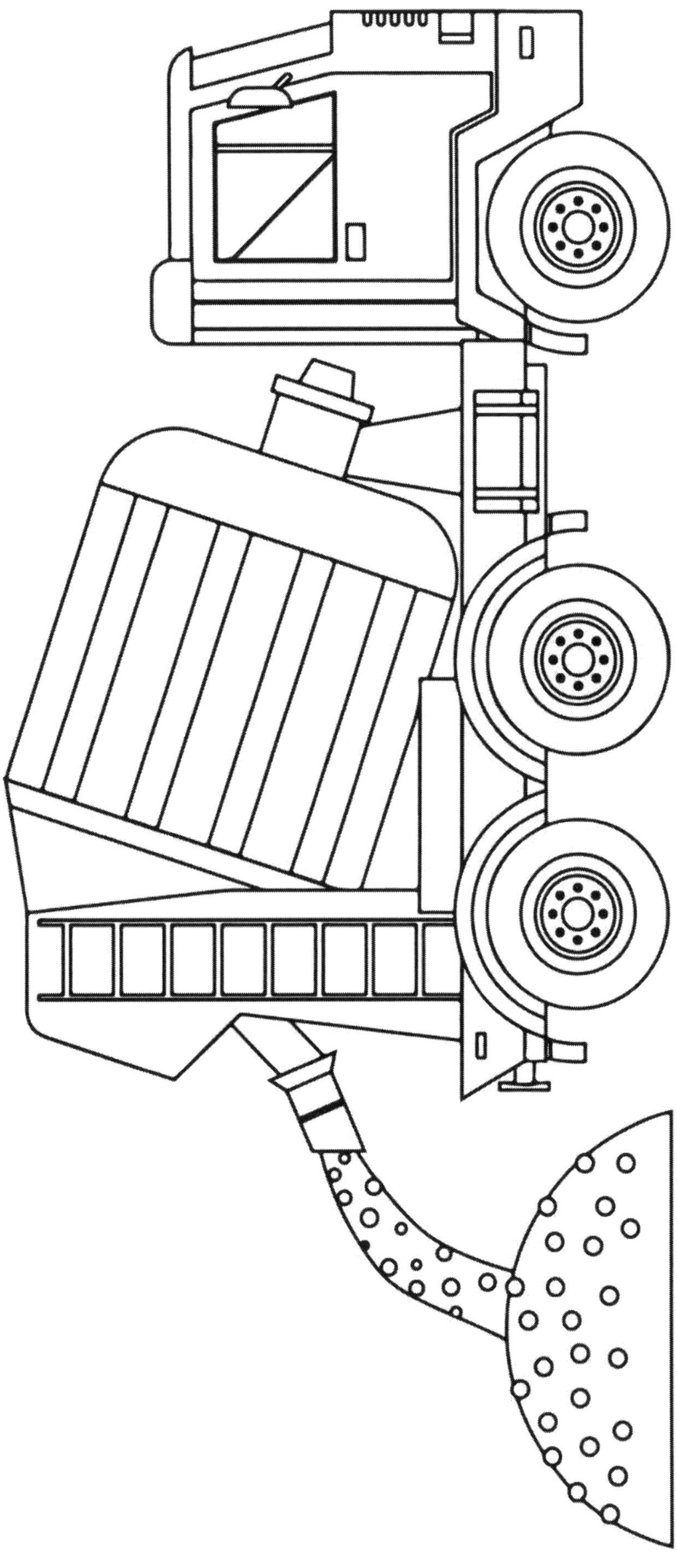

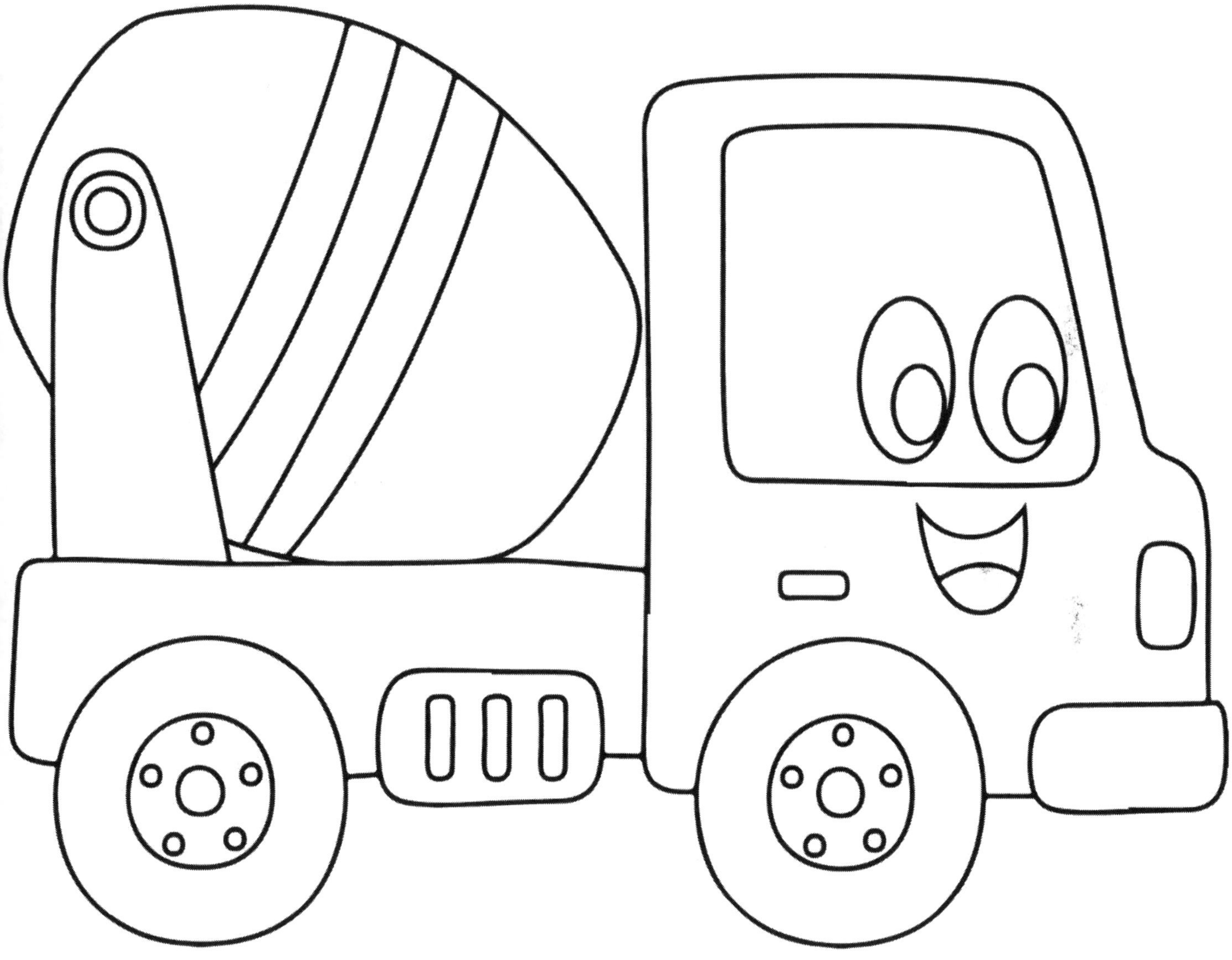

Made in the USA
Las Vegas, NV
21 June 2021